Signal Integrity Issues and Printed Circuit Board Design

Table of Contents	xiii

    Slots in Planes 3 232
    Guard Bands 233

13 Crosstalk Simulations 235

    Basic Model 235
    Add an Uncoupled Region 239
    Effect of Length 240
    Stripline 242
    Stripline With Terminations 242
    More Realistic Example 244
    Summary 247

14 Differential Traces and Impedance 249

    Background 249
        Advantages 249
        Key Assumptions 251
    Design Rules 255
        Design Rule 1 (Equal Length) 255
            Common Mode Implications 255
            Differential Signals and Loop Areas 256
        Design Rule 2 (Close Together) 257
            Rule 2 Consequence 258
        Design Rule 3 (Differential Impedance) 260
            Differential Mode Impedance 260
            Common Mode Impedance 261
        Design Rule 4 (Constant Separation) 262
    Differential Simulations 262
    Calculating Differential Impedance 265
        Edge Coupled 266
        Broadside Coupled 268

15 Bypass Caps and Decoupling Systems 269

    Traditional Approach 269
        Size and Quantity 273
        Placement 275
        Connection 276
    Power System Impedance Approach 277
        Ideal Response 277
        Capacitor Response 278
        Inductance Effects 278
        Multiple Capacitors 280
        Additional Capacitor Value 281

Planar Capacitance	282
Equivalent Series Resistance (ESR)	285
Self Resonant Frequencies	287
Effects of Multiple Capacitors	288
Parallel Capacitors	288
Resonance	290
Impedance at Self-Resonant Frequency	291
Impedance at Anti-Resonance	292
General Case Analysis	293
Consequences	294
Summary	296

16 Power Systems 297

Power Supply Voltages	297
Need For Power Planes	300
Strategies For Designing With Plane	300
Some Design Rules	303
Connecting Reference Planes Together	303
Overlapping Planes	304
Decoupling to Wrong Plane	305
Signals Crossing Separations	305
Stack-Ups	307
Conclusion	308

17 Lossy Lines and Eye Diagrams 311

Lossy Lines	311
Skin Effect	312
Dielectric Absorption	314
Lossy Line Model	314
Eye Diagrams	316
Equalization	318
Passive Equalization	319
Active Equalization	319
Summary	320

Part 3 Appendices, Glossary, and Index 323

Appendices 325

A	UltraCAD's Square Wave Simulator	325
B	Why Inductors Induct	329
C	Logarithms	345

Signal Integrity Issues and Printed Circuit Board Design

Douglas Brooks

Prentice Hall PTR
Upper Saddle River, NJ 07458
www.phptr.com

Library of Congress Cataloging-in-Publication Data

A CIP catalog record of this book can be obtained from the Library of Congress

Production Supervisor: Wil Mara
Publisher: Bernard M. Goodwin
Cover Design: Nina Scuderi
Cover Design Director: Jerry Votta
Editorial Assistant: Michelle Vincenti
Marketing Manager: Dan DePasquale
Buyer: Maura Zaldivar
Composition: Douglas Brooks

© 2003 Pearson Education, Inc.
Publishing as Prentice Hall Professional Technical Reference
Upper Saddle River, New Jersey 07458

Prentice Hall PTR offers excellent discounts on this book when ordered in quantity for bulk purchases or special sales. For more information, please contact: U.S. Corporate and Government Sales, 1-800-382-3419, corpsales@pearsontechgroup.com. For sales outside of the U.S., please contact: International Sales, 1-317-581-3793, international@pearsontechgroup.com.

Company and product names mentioned herein are the trademarks or registered trademarks of their respective owners.

All rights reserved. No part of this book may be reproduced, in any form or by any means, without permission in writing from the publisher.

Printed in the United States of America

Second Printing

ISBN 0-13-141884-X

Pearson Education Ltd.
Pearson Education Australia Pty., Limited
Pearson Education Singapore, Pte. Ltd.
Pearson Education North Asia Ltd.
Pearson Education Canada, Ltd.
Pearson Educación de Mexico, S.A. de C.V.
Pearson Education—Japan
Pearson Education Malaysia, Pte. Ltd.

DEDICATION

There are a great many PCB designers in the world who do not have a technical degree. They struggle each day to keep up in an industry that is advancing very rapidly. This book is dedicated to them with the sincere hope that it will help them in their journey.

About the Web Site

The author is president of UltraCAD Design, Inc. UltraCAD maintains a Web site at *http://www.ultracad.com* where you can find a wealth of information about board design.

The site includes—

- Reprints of the articles the author has written during the past seven years
- Technical and design notes related to the board-design process
- Five freeware calculators to help board designers handle difficult tasks
- Information regarding seminars Brooks offers from time to time

As support/errata information becomes available for the book it will be posted on this site.

Prentice Hall Modern Semiconductor Design Series

James R. Armstrong and F. Gail Gray
   VHDL Design Representation and Synthesis

Mark Gordon Arnold
   Verilog Digital Computer Design: Algorithms into Hardware

Jayaram Bhasker
   A VHDL Primer, Third Edition

Eric Bogatin
   Signal Integrity: Simplified

Douglas Brooks
   Signal Integrity Issues and Printed Circuit Board Design

Kanad Chakraborty and Pinaki Mazumder
   Fault-Tolerance and Reliability Techniques for High-Density Random-Access Memories

Ken Coffman
   Real World FPGA Design with Verilog

Alfred Crouch
   Design-for-Test for Digital IC's and Embedded Core Systems

Daniel P. Foty
   MOSFET Modeling with SPICE: Principles and Practice

Nigel Horspool and Peter Gorman
   The ASIC Handbook

Howard Johnson and Martin Graham
   High-Speed Digital Design: A Handbook of Black Magic

Howard Johnson and Martin Graham
   High-Speed Signal Propagation: Advanced Black Magic

Pinaki Mazumder and Elizabeth Rudnick
 Genetic Algorithms for VLSI Design, Layout, and Test Automation

Farzad Nekoogar and Faranak Nekoogar
 From ASICs to SOCs: A Practical Approach

Farzad Nekoogar
 Timing Verification of Application-Specific Integrated Circuits (ASICs)

David Pellerin and Douglas Taylor
 VHDL Made Easy!

Samir S. Rofail and Kiat-Seng Yeo
 Low-Voltage Low-Power Digital BiCMOS Circuits: Circuit Design, Comparative Study, and Sensitivity Analysis

Frank Scarpino
 VHDL and AHDL Digital System Implementation

Wayne Wolf
 Modern VLSI Design: System-on-Chip Design, Third Edition

Kiat-Seng Yeo, Samir S. Rofail, and Wang-Ling Goh
 CMOS/BiCMOS ULSI: Low Voltage, Low Power

Brian Young
 Digital Signal Integrity: Modeling and Simulation with Interconnects and Packages

Bob Zeidman
 Verilog Designer's Library

TABLE OF CONTENTS

PREFACE xvii

Part 1 BASIC CONCEPTS 1

1 Electronic Concepts 3
    Current 3
    Charge 5
    Voltage 6
    Direct and Alternating Voltage and Current 8
    Harmonics 11
    Measurement of AC Voltage or Current 12
    Frequency, Rise/Fall Times, and Period 13
    Frequency Measurement 16
    Complex Waveforms (Fourier Analyses) 18
    Chapter Endnotes 23
        Direction of Current Flow 23
        Some Standard Symbols Used in Electronics 24
        Some Fundamental Definitions 25
        Impportant Names in Electronics 26

2 Propagation Times 27
    Propagation Speed 27
    Propagation Times 28
    Trace Configurations and Signal Propagation 29
    Circuit Timing Issues 32
    Wavelength 35

3 Electrical Components 37

The Three Basics 37
Resistance 38
Ohm's Law 39
Capacitance 43
Charge Storage 47
Formula for Capacitance 48
Capacitance Functions and Effects 49
Inductance 50
Formulas for Inductance 55
Charge and Discharge Currents 56
Resonance 60

4 Voltage and Current Changes and Time Constants 63

Voltage and Current Changes Through Resistors 63
Voltage and Current Changes Through Capacitors 64
Voltage and Current Changes Through Inductors 66
Some Interesting Inductive Circuit Dynamics 69
Time Constants 71
A Note on Charge and Discharge Equations 76

5 Resistance 77

Kirchhoff's Laws 77
Series Resistors 79
Parallel Resistors 81
Voltage Dividers 83
Amplifier Feedback and Gain 85
Power 86
Equivalent Circuits 88
Power Curve 91
Power Sources 93
Conductance 94

6 Reactance 97

Capacitive Reactance 98
Inductive Reactance 99
Ohm's Law for Reactance 101
Series LC Combinations 104
Parallel LC Circuits 107
Resonance 111
Poles and Zeros 115

Table of Contents

Susceptance 119

7 Impedance and Phase Shift 121

Impedance 121
Effect of Frequency 128
Another RC Example 129
Classic RC Filter 131
Combining Impedances 133
Resonance and Q 134
Series RLC Circuits 137
Series RLC at Resonance 140
Admittance 142
Chapter Endnotes 144
    Detailed Calculations for Figure 7-8 144
    RLC Simulation 145
    Imaginary Numbers 146

Part 2 SIGNAL INTEGRITY ISSUES 147

8 Signal Integrity Overview 149

9 Electromagnetic Interference (EMI) 155

Background 155
Fields and Cancellations 156
Some Basic Truths 158
Signal Coupling 159
Loop Area 161
    Slots in Planes 1 162
    Return Pathways 163
    Power Plane Returns 164
    Changing Trace Layers 165
    Unrelated Planes 166
    Stripline 167
Stubs 168
Common Mode 169
The 20-H Rule 173
Picket Fences (Faraday Shields) 174

10 Reflections and Transmission Lines 175

Communications Model 175
Transmission Lines 178
Critical Length 181
Reflection Coefficients 183
Visualizing Reflections 184
Determining Trace Impedance 185
Termination Techniques 190
    Parallel 192
    Thevenin 192
    AC Termination 194
    Series 194
    Diode 195
Some Design Issues 196
    Changing Trace Layers 1 196
    Power Planes 197
    Changing Trace Layers 2 197
    Slots in Planes 2 199
Stubs 200
Absolute vs. Relative Value of Zo 201
Chapter Endnote 203
    On Formulas 203

11 Some Transmission Line Simulations 207

Basic Simulation 207
Series Termination 215
Placement Issues 217
Branches, or Ys 219

12 Crosstalk 221

Forward versus Backward Crosstalk 221
    Backward Crosstalk 223
    Forward Crosstalk 225
    In Summary 226
Estimating Crosstalk 226
    Calculations 226
    Coupling 227
    Distance 227
    Terminations 229
    UltraCAD Calculator 230
    HyperLynx Simulation Tool 231
Design Considerations 232

D	Phase Shift Simulation	351
E	Complex Algebra	357
F	Transmission Line Simulation	367
G	Echo Illustration	369
H	UltraCAD Freeware Calculators	371
I	TDRs and VNAs	375
J	Right Angle Corners	383

Glossary 387

Index 397

About the Author 403

PREFACE

HOW IT ALL STARTED

*W*riting has been an avocation of mine for most of my career. And I have spent a few of my years teaching at the college level. After I became involved in the printed circuit board (PCB) design business in 1991 I submitted a couple of articles, which were subsequently published, to *Printed Circuit Design* magazine. Later, Pete Waddell asked me, as one of his contributors, to come to a press function at PCB West in 1997. Afterwards, I half-facetiously asked him if he would send me to PCB East the following fall. His answer was, "Sure, if you'll put on a seminar."

And that was how it all started.

I presented a seminar that fall entitled "Circuit Noise and EMI Control." My motivation was twofold: First I wanted to make a contribution back to an industry that had been good to me through the years. And, I wanted to go to Boston, a city I have always enjoyed visiting. Soon, I had several different seminars I was offering.

There are several of us who are sometimes considered "regulars" on the seminar circuit. Most of us like each other, respect each other, sit in on each other's seminars, and learn from each other. After a couple of years I realized that all of our collective signal integrity seminars talked about things like capacitive and inductive coupling and bypass capacitor lead-inductance-caused antiresonant impedance peaks, and so on, but most of the seminar attendees had no technical background. They didn't know what a capacitor was, much less what an antiresonant peak was!

That led to the development of a one-day "course" in electrical engineering. We titled it "Electrical Engineering for the Non-Degreed Engineer," a title that varies from time to time. I wanted to add the words "and for the engineer who didn't get it the first time" to the title, but Pete wouldn't let me (probably for good reason). I don't know how to politely say this, but I know there are a large number of engineers who just didn't get the EMI and magnetic coupling stuff when it was presented in school, and almost no college program addresses the practical engineering problems associated with board design in their college-level curriculums. There are more people who need to read this book than we can politely invite!

Occasionally I am asked if we need to keep putting on these seminars and writing these books. After all, with more and more electronics being packaged in the chip, won't the board problems disappear? Won't higher and higher levels of circuit integration make PCBs obsolete?

I love that question. I heard it the first time when the integrated circuit was developed. I heard it again when the microprocessor was developed. And we hear it every time there is another increase in system-level integration. My answer is simple. Printed circuit boards are not going to diminish in importance in the future for the same reason that they haven't diminished in importance in the past. There will *always* be a need to interconnect devices on some sort of physical surface.

ACKNOWLEDGMENTS

There are a great many people who have contributed to this book in some way or another. Pete Waddell and his team have been great to work with through the years on both the article and the seminar levels. I have been very grateful to have met and come to know many of the leaders in our field—wonderful colleagues who learn from each other and who have a lot of fun together. People like Glenn Wells and Gary Farrari have been inspirations as I have watched how much they have given back to the development and growth of new designers starting out in our field. Glenn, in particular, seems to be everywhere in our industry—classroom, seminar room, college administrator's office pounding the desk for more funds for program development, Top Gun contestant, and Top Gun program coordination. I am honored to count him as a friend and I appreciate the help he has given me with this book during the various phases of its development.

My partner, Dave Graves (himself a Top Gun award winner) has been invaluable in his review of my seminar materials and article drafts as they have been prepared through the years. He has often been the first to see something as it was being created, and I can't tell you how many times I've gone back to the drawing board when his response was "I don't get this! What are you trying to say?" Everything is better because of his reviews and support (and numerous contributions).

I would also like to acknowledge three vendors who have been especially generous with their support for my article and seminar activities through the years. Mentor Graphics, HyperLynx (now a part of Mentor), and Polar Instruments have offered licensing and technical support whenever asked. And I appreciate the fact that they have offered that support without ever trying to exert any type of control over any of the ways that support was being used.

I've enjoyed preparing this book and have mixed emotions now that it is finished. But there are a couple of people who are *really* happy this project is finally finished, particularly my (nontechnical) wife. She has been wondering what all this fuss has been about, but is glad that I finally can think and talk about other things.

Finally, I am grateful that Bernard Goodwin at Prentice Hall decided to publish this. Going through the process of preparing a manuscript for formal publication with Bernard and with Wil Mara has been a real learning experience for me. I hope you, the reader, determine that it has all been worthwhile.

PART 1

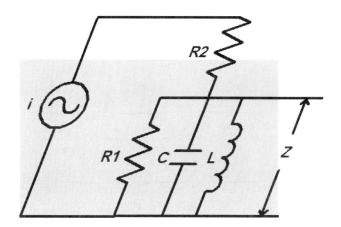

BASIC CONCEPTS

Charles Augustin de Coulomb is credited with Coulomb's Law (1785) regarding positive and negative charges. In the century between that and the publication of James Clerk Maxwell's Equations in 1873, the knowledge of electronics expanded rapidly. This Part 1 covers the basic concepts of electronics. It starts with perhaps our most fundamental definition: Current is the flow of electrons. It ends with impedance as a complex function with both magnitude and phase shift. Along the way we discuss many fundamentals: the nature of resistance, capacitance, and inductance; combinations of these components; fundamentals of waveforms; simple circuits; and time constants. The objective is not to turn you into an engineer. This part is not that complete. But the objective is to give you the fundamental background you will need to understand the concepts discussed in Part 2.

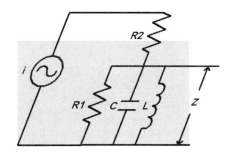

1 ELECTRONIC CONCEPTS

CURRENT

Current is the flow of electrons.

That is a great sentence. But its simplicity belies its importance and significance. Almost every fundamental definition in electronics is based on this one, and almost every problem in signal integrity has its basis in the flow of electrons, especially if that flow changes rapidly. If I were to identify the two most important concepts in this book, this would be one of them. (The other would be Ohm's Law.)

So, let's look at this thing called an *electron*. In the early 1900s a model of the atom was developed that consisted of a nucleus, containing protons and neutrons, and one or more outer shells containing electrons (Figure 1-1). It is called a *planetary* model because it resembles a central star with planets circling around it. The protons are positively charged and the electrons are negatively charged. Since the net charge must (in a stable state) be zero, there must be an equal number of protons and electrons in any atom. The atomic number of an atom represents the number of protons in the nucleus (which also represents the number of electrons), and the atomic weight of an atom is the sum of the protons and neutrons in the nucleus.

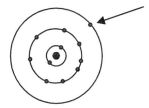

Figure 1-1 Planetary model of an atom.

We picture electrons as circling in shells, or orbits, around the nucleus. Shells have a maximum number of electrons they can hold. The innermost shell, for example, can only contain a maximum of two electrons. As one shell, or orbit, fills, electrons start occupying the next shell. (In fact, these shells are more like energy states than physical positions, but we don't know how to draw pictures of energy states.) Thus, hydrogen, the most basic element, has one proton and one electron. It has an atomic number of one. Its electron exists in the innermost shell. Helium has two protons, two electrons, and an atomic number of two. Its electrons completely fill the innermost shell. Lithium, with an atomic number of three, has three protons and three electrons. Two of those electrons occupy the innermost shell, and the third one occupies the next higher shell.

If you are familiar with the Periodic Table of the Elements, you are aware that one way elements are grouped is by the number of electrons in an element's outer shell. We also know intuitively that gold, silver, and copper are among the very best conductors of electricity. We might ask what makes these three elements so special. Well it happens that among the things these three elements have in common is that they (a) are metals, (b) they are solids at normal temperatures, and (c) they each have only a single electron in their outermost shell.

We can think of a single electron in the outermost shell as being loosely held by the nucleus and being easily dislodged by external forces. Remember the rule that like charges repel and unlike charges attract. If an external electron were to somehow approach an atom with a single electron in the outermost shell, it might easily dislodge and repel the atom's outer electron and replace it.

Now, let's say we had a whole bunch of copper atoms, stretched out in a wire. We "put" an electron into one end of the wire (perhaps by using a battery). That electron might dislodge the outer electron belonging to the first atom and replace it. That atom's electron, now free, might move along and replace the next atom's outer electron, and so on. When we finally get to the last copper atom in the wire, that atom's electron must have a place to go if it is to be replaced. If we have connected a battery at both ends of the wire, this last electron will flow back into the battery. If we have not connected the battery to both ends of the wire, there is no place for the last electron to go. Since there is no "room" for it, the process will immediately stop and no (current) flow occurs.

This very simple illustration already highlights three points:

Chapter 1

1. This movement of electrons constitutes current.
2. Current must have a place to go; that is, it must flow in a loop. If there is no connection at the far end of the wire, there is no place for the last electron to go. Electrons will instantly build up and block the flow of any additional electrons into the wire. Electrons can only *flow* along the wire if there is a connection at both ends of the wire.
3. Batteries consist of a chemical process that provides a source of electrons at one terminal, and a corresponding shortage of electrons at the other terminal. Thus, a battery can *push* an electron in one end of a wire, as long as it receives an electron at its other terminal.

A common misconception is that the electron that goes in one end of a wire is the very same one that comes out the other end. As the illustration above suggests, this is not true. It may take a very long time (even hours) for an individual electron to travel through a wire. As an analogy, picture a *very* long train tunnel filled with identical cars moving through it. As one car enters the tunnel, another (identical) car leaves. There is a flow rate we can measure in terms of cars-per-unit-time; each time one car goes in, one car goes out. But it may take a very long time for the individual car that just went in to finally emerge at the other end.

In truth, neither illustration is exactly correct. The electrons do not move from atom to atom. They may skip over several atoms before they dislodge and replace another electron (see Figure 1-2). In crystalline copper, the overall average length of travel of an individual electron is approximately four atoms.

CHARGE

The fundamental unit of measure of charge is the coulomb (named for Charles Augustin de Coulomb. (See the note at the end of this chapter for a list of names and lifetimes of significant people in the development of electronics.) One coulomb is defined as the combined charge of 6.25×10^{18} electrons. One amp of current (named for Andre Marie Ampere) is defined as the flow of one coulomb of charge in one second of time. Thus, a current of one amp is the flow of 6.25×10^{18} electrons passing one point in one second. This is the core definition in electronics. Everything else is derived from it. (See the summary of definitions at the end of this chapter.)

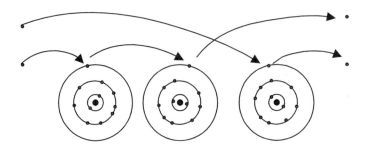

Figure 1-2 Electrons can dislodge and replace those in adjacent atoms.

Charge has polarity. Electrons have negative charge. Protons have positive charge. But charge is also a relative concept. Another way to have positive charge is to have a shortage of, or fewer, electrons. (We will later learn that as electrons flow onto one plate of a capacitor, electrons flow away from the other plate. Thus one plate has a positive charge *relative* to the other one. But both plates could be negatively charged with respect to a third plate with still fewer electrons.) Coulomb's Law states that like charges repel each other and unlike charges attract each other, with a force that is inversely related to the square of the distance between them. Two electrons, for example, will repel each other.

VOLTAGE

Voltage is a force. It is the force of voltage that causes electrons to flow (current). It is easiest to think of voltage as being a buildup of charge. For example, if we had a bunched-up "ball" of electrons, each with a negative charge, they would generate a force that would try to move themselves away from each other (since like charges repel each other). The magnitude of this force (the voltage) would be directly related to the number of electrons bunched together.

We often use hydraulic analogs to help us visualize voltage and current. Picture water flowing through an ordinary garden hose. It will flow out just a little way from the end of the hose. That is analogous to current flowing through a wire. There is only a little pressure built up in the hose, mostly caused by friction. The bigger the hose, the smaller the friction, and the closer the water will fall to the end of the hose.

Now put your thumb across half the end of the hose. You are impeding (as in impedance) the flow of the water. Pressure builds up against your thumb. This is analogous to voltage building up. The impedance causes a smaller flow of water (smaller current), but the pressure increases and causes the water to squirt further away from the end of the hose.

Using your thumb, now cover three-fourths of the end of the hose. A smaller amount of water (current) flows, but it squirts out further because of the increased pressure (voltage). If you tried to cover the end of the hose completely with your thumb, you probably couldn't. Few of us have enough strength in our hands and thumbs to completely counteract the pressure of the water.

This example intuitively points out that there is a relationship between impedance (your thumb), current (water flow), and voltage (pressure). Increasing the impedance results in less current and/or higher voltage.

Voltage is the "push" behind current. There are only a few ways we can create voltage sources. One way is to build a supply of electrons on a cathode and a "shortage" of electrons on an anode of a battery through chemical action. The difference in charge creates the force we call voltage. If we connect a circuit between the anode and cathode of a battery, current will try to flow through the circuit. Another way to create voltage is to use changing magnetic energy to "generate" the force to move electrons (the principle behind a generator or a transformer). We can temporarily store energy (voltage) in something like a capacitor. Fuel cells, photocells, and thermocouples can convert other forms of energy to voltage or current. And we can accidentally generate voltage simply by walking across a carpet (static electricity.)

We generally think of "high" voltage as being dangerous and "low" voltage as being safe. But this is not always true. Sometimes at fairs or magic shows there is equipment that will give us a one-million-volt shock without harm! And you can seriously burn yourself around a 12-volt car battery if you accidentally short the two terminals. It is power, the combined effect of voltage and current, that is dangerous, not necessarily just voltage by itself.

Direct and Alternating Voltage and Current

We generally think of current and voltage as either direct current (DC) or alternating current (AC). DC voltage holds a steady value and DC current flows at a constant rate in a single, constant direction, at least during the relevant time of consideration. AC voltage and current change with time. We usually think of these changes as going in one direction, + (positive) voltage or current, to the opposite direction, − (negative) voltage or current, and then going back in the first direction again. But that does not always have to be the case. AC voltage could, for example, simply change from one positive level to another positive level. We also tend to consider these changes as being periodic, or repetitive. But the changes could also be random. (See the note at the end of this chapter about which direction the electrons flow.)

Figure 1-3 illustrates two typical classic AC waveforms. These are sine waves and they are fundamental to the analysis of what happens in electronic circuits. The signal grows to a positive peak and then starts to decline in magnitude until it crosses over the horizontal axis. At that time, the signal reverses itself and becomes negative. It grows to a negative peak and then declines in (negative) amplitude until it crosses the horizontal axis and reverses itself again. It continues in this fashion indefinitely.

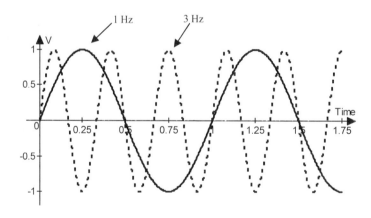

Figure 1-3 Two typical voltage sine waves.

The formula for a sinusoidal voltage waveform in its simplest form is

$$V = A \times \sin(360 \times f \times t)$$

where A is the amplitude of the waveform, 360 is the number of degrees in a cycle, f is the frequency of the waveform in Hertz (cycles per second), and t is time in seconds.

An AC waveform is characterized by three measures, its frequency (f), its amplitude (A), and its phase (Θ). Thus, a more general description of a simple voltage waveform might look like this:

$$V = A \times \sin(360 \times f \times t + \Theta)$$

Perhaps you recall from high school geometry that a sine wave and a cosine wave are shifted from each other by 90 degrees. That is, $\sin(x) = \cos(x - 90)$, or $\sin(x + 90) = \cos(x)$. We say a cosine waveform is shifted 90 degrees *ahead* of a sine waveform (the sine wave doesn't reach the same value as the cosine wave until 90 degrees later). In this instance we say that the cosine waveform *leads* the sine waveform by 90 degrees or that the sine wave *lags* the cosine wave by 90 degrees (see Figure 1-4). In this illustration, the phase shift, Θ, between a sine and a cosine waveform is 90 degrees. Phase shift is an important concept in impedance, as we will see in Chapter 7.

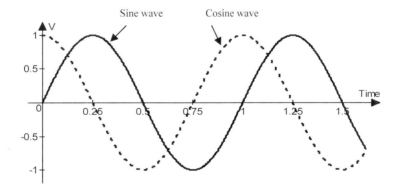

Figure 1-4 A 1-Hz cosine wave leads a 1-Hz sine wave by 90 degrees.

Phase shift, Θ, is relative. So we always need to know what the arbitrary reference point is when we are using formulas like this. Often, but not always, the reference phase (where Θ = 0) is that of the driving voltage or driving current signal in our circuits. Then we speak about other voltages and currents being shifted in phase from this reference.

A sine wave is symmetrical around the horizontal axis. So, although it has a peak value (equal to A = 1.0 in Figure 1-4) and a peak-to-peak value (equal to 2.0, from −1.0 to +1.0, in Figure 1-4), its average value through time is zero.

A sinusoidal (really trigonometric) waveform is a naturally occurring waveform in nature. A bouncing ball can be described by such a waveform. Oceanic tidal action is described by a combination of such waveforms. The motion of planets around stars is described by such waveforms. The flow of AC currents is described by such waveforms. And the mathematics of calculus deals nicely with such waveforms. The problem is, the shapes of the signals in our circuits are usually not described by such waveforms (at least not directly). For example, a clock square wave doesn't look very "trigonometric" and a simple audio waveform, Figure 1-5, is decidedly not periodic!

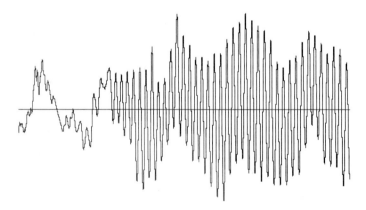

Figure 1-5 Picture of a nonperiodic audio waveform.

HARMONICS

Let's say we have a signal with a certain frequency and that we can describe that waveform by the relationship

$$V = \sin(x)$$

where x is a measure of frequency.

Then a waveform whose shape is given by the formula $V = \sin(2x)$ has twice the frequency of the first waveform, and another waveform $V = \sin(3x)$ would have three times the frequency. Figure 1-6 illustrates a signal and its third harmonic (reduced in amplitude).

We can generalize this by saying that a waveform with shape $V = \sin(nx)$ has a frequency n times higher than a waveform whose shape is $V = \sin(x)$. If the waveforms are related to each other [perhaps the $\sin(nx)$ waveform was somehow generated by the $\sin(x)$ waveform, or perhaps they were both generated by the same source] then such shapes (waveforms) are called harmonics. Harmonics are simply multiples of the base, or fundamental, frequency. In this case the $\sin(nx)$ signal is called the nth harmonic of the fundamental $\sin(x)$ signal. Harmonics will become very important later.

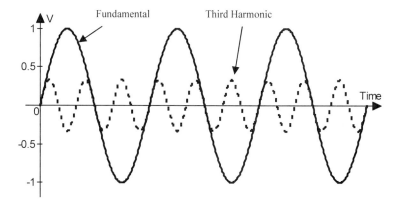

Figure 1-6 A fundamental waveform and its third harmonic.

MEASUREMENT OF AC VOLTAGE OR CURRENT

The measurement of DC voltage or current is straightforward. It is simply the observed value. But the measurement of AC waveforms is a different matter. Assume the waveform cycles around zero (the horizontal axis, see Figure 1-7). We can't use the average value, since that is always zero no matter what the peak value is. The peak value (1.0 V in Figure 1-7) and the peak-to-peak value (2.0 V in the figure) also are misleading, since that doesn't take into consideration the "shape" of the waveform.

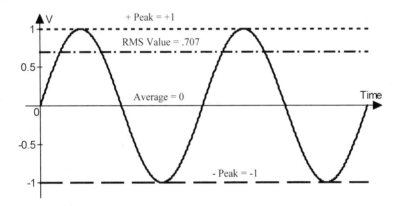

Figure 1-7 RMS value in relationship to other measures of amplitude.

Instead, we use a measurement value for AC waveforms called the RMS value. RMS stands for Root-Mean-Square. Conceptually, we find the RMS value by breaking up the waveform into a large number of small sections, and then (a) squaring the amplitude of the waveform for each section, (b) calculating the mean value of all of these squared vales, and (c) then taking the square root of this mean value. Mathematically we represent the RMS value of a waveform with this formula:

$$\text{RMS} = \sqrt{\frac{1}{n} \sum_{1}^{n} \sin^2(n)} \qquad [1\text{-}1]$$

The RMS value of a 1.0-V sine wave as shown in Figure 1-7 works out to be .707 volts. In general, the RMS value of a sine or cosine waveform is .707 times the peak value. That means that the 117-volt AC waveform in our homes has a peak value of 117/.707 = 165.5 volts and a peak-to-peak value of 231 volts.

The true RMS value of any waveform actually relates to power. Suppose we applied the sine wave voltage waveform in Figure 1-7 across a resistor. The RMS value is the equivalent DC voltage that would generate the same power (i.e., heat) in the resistor. Thus, a DC voltage of .707 V across a resistor will heat the resistor exactly the same amount that a ± 1.0 V sine wave would.

But this value, .707, is only the true RMS value for a sine wave. What if the waveform were a different shape? What if the waveform were random? The audio waveform shown in Figure 1-5 seems pretty random when you look closely at it. How could we then measure its RMS value?

A simple AC meter measures AC waveforms accurately only if they are sine waves. The meters typically rectify the AC waveform and are then calibrated to apply a fraction to the peak value or to the average value of the rectified waveform. But this technique only works when the shape of the waveform is known to be sinusoidal. There are really only two primary ways to measure the "true" RMS value of a nonsinusoidal waveform. In the past, so-called "true" RMS meters often applied the test voltage across a resistor and measured the heat generated, inferring the RMS value that way. More recently, with improved calculational power, we can use microprocessor-based instruments to actually do the calculations in Equation 1-1.

Even today, meters that do not *expressly* indicate that they are true RMS meters probably only measure AC voltages accurately if the waveforms are sinusoidal.

FREQUENCY, RISE/FALL TIMES, AND PERIOD

Frequency is the measure of how many times the current cycles through a change in direction in one unit of time (usually one second). Thus, the dotted curve in Figure 1-3 illustrates a higher frequency wavelength than does the solid line waveform in that figure. If the horizontal axis in the figure represents seconds, then the solid waveform in the figure has a frequency of one cycle per second (or 1 Hertz) and the dotted waveform has a frequency of 3 Hertz. We now commonly see waveforms on our circuit boards with frequencies in the range of hundreds of millions of cycles each second (hundreds of MHz) and sometimes much higher. That means the AC waveform, or the current, is cycling back and forth, or changing direction, hundreds of millions of times each second!

It is a common misconception that *frequency* is the issue in high-speed designs. As we will learn in this book, it is *rise time* that is the real culprit. Figure 1-8 illustrates two waveforms, one a sine wave and the other a square wave. (The square wave is representative of what we would like (even wish) our clock signals to resemble) Each of these waveforms has the same frequency. But these two waveforms have significantly different rise times.

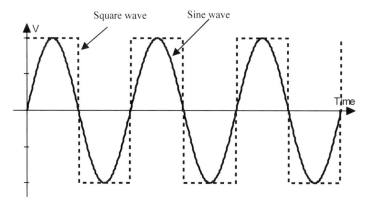

Figure 1-8 Frequency versus rise time.

Rise time is normally defined as the length of time required for the waveform to rise from the 10% point on the waveform to the 90% point (Figure 1-9). Occasionally you will see rise time defined between the 20% and 80% points. Fall time is defined exactly the same way, the time for the waveform to fall from the 90% point to the 10% point. If you superimposed a sine wave alongside the rising edge of a signal, as shown in Figure 1-9, it can be shown that the 10% to 90% portion of a sinusoidal waveform is just about one-third the time for a complete cycle, or period, τ, of the waveform.

The time required for a single cycle is 1/f where f is the frequency. The time for a single cycle is called the period (τ). So a sine wave with a frequency of 1 MHz (one million cycles in a second) has a period of one-millionth of a second,

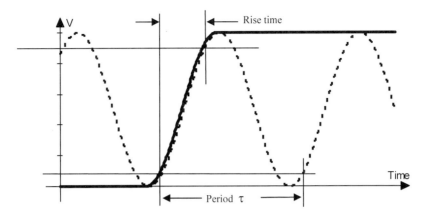

Figure 1-9 The rise time of a pulse is approximately 30% of the period of the underlying sine wave.

or 1 µs, or 1,000 ns. Therefore, the rise time of a 1-MHz sine wave is about one-third of that, or about 333 ns.

You might ask why we refer to the 10% to 90% range of amplitude for rise time and not the 0% to 100%. Or why some use the 20% to 80% measure for rise time. The answer may be somewhat less than satisfying, but it goes like this: Almost all switching devices follow a switching pattern similar to that shown by the rising line in Figure 1-9, but they may differ significantly in the very first or last portions of their switching pattern. For example, one type of device may reach 100% very quickly, but another device may roll off as it approaches the 100% level, and actually may not reach 100% for several moments (we sometimes say it may approach 100% *asymptotically*). Switching devices might have quite different switching patterns in the very beginning or ending stages of their switching range, but usually all have quite similar patterns during their main transition. Therefore, we use the 10% to 90% points of the switching range for the rise time measure in order to compare "apples to apples." Sometimes others may use the 20% to 80% range because the beginning and ending parts of the switching range are a little "broader" than other devices, or because they are engaging in "specsmanship" (the 20% to 80% range is shorter and therefore faster).

Suppose we have a voltage or current requirement in a circuit that changes rapidly. For example, suppose we have a current requirement that goes from 0 mA to 10 mA within 1 ns. We can express this requirement as a "change in current divided by a change in time," or $\Delta i/\Delta t$ (where Δ means change). If we consider the relationship, $\Delta i/\Delta t$, and think of Δt being an extremely short period of time, then we would mathematically express the relationship as di/dt (pronounced dee-eye-dee-tee). The term di/dt is a calculus expression that means "the change in current divided by the change in time as the change in time becomes vanishingly small,"— that is, as Δt becomes really small! In really fast circuits, the term dt equates to the rise (or fall) time of the signal. We will see later that it is this di/dt term that causes the signal integrity problems.

We in the industry often simply use the term *rise time* generically to describe a circuit requirement. The reader should understand that fall time is equally important. What is really important is the faster of the two. So when you read "rise time" think "rise or fall time, whichever is faster."

FREQUENCY MEASUREMENT

We actually can measure, or refer to, frequency in three ways. The most common one is in cycles per second, or Hertz, often designated by the symbol "f." Thus, the waveform shown in Figure 1-10 has a frequency of 3 Hertz, or f = 3. This is a very common measure of frequency. We can also refer to the number of degrees the waveform goes through in one second. A sine wave completes 360 degrees in one cycle, so in three cycles it completes 360 x 3 or 1080 degrees. Thus, the waveform in Figure 1-10 could be said to have a frequency of 360 x f or 1080 degrees/second. In general we could describe any frequency as 360 x f degrees/second. We almost never use this measure of frequency!

But there is a related measure of angular frequency we *do* use. We start by dividing the circumference of a circle into radians. A radian is the angle formed by a length along the circumference of a circle equal to the radius of the circle (Figure 1-11). Take a length equal to one radius and lay it along the circumference. Look at the angle formed by the beginning and the end of this length (arc). That angle is defined as a radian.

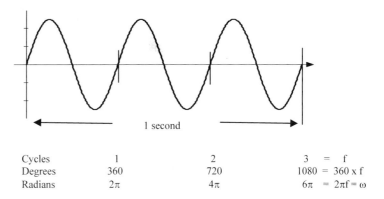

Cycles	1	2	3	= f
Degrees	360	720	1080	= 360 x f
Radians	2π	4π	6π	= $2\pi f$ = ω

Figure 1-10 Alternative ways of measuring the frequency of a waveform.

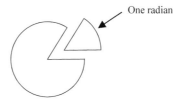

Figure 1-11 Definition of a radian.

The circumference of a circle is $2\pi r$, where r is the radius. So if we ask how many radians there are in a circle (360 degrees) the answer is:

$$\text{Circumference/radian} = 2\pi r/r = 2\pi.$$

So, 2π radians make up a complete circle (360 degrees).

Since there are 2π radians in 360 degrees, there are therefore $2\pi f$ radians completed by a sine wave in one second. This measure of frequency we *do* use in electronics and it is given a special symbol, ω, as shown in Equation 1-2.

$$\omega = 2\pi f \qquad [1\text{-}2]$$

This measure of frequency is called the *angular frequency* and shows up in a great many formulas used in electronics. It measures the number of radians a sine wave goes through (completes) in one second.

COMPLEX WAVEFORMS (FOURIER ANALYSES)

We will find later in this book that there are only three types of passive components we have to worry about on our boards and in our designs, resistors, capacitors, and inductors. Each type of component offers an *impedance* to the flow of current. For a resistor, this impedance is simply a matter of the value of the resistor's size. But for capacitors and inductors, the impedance is a function of both the component's size and the frequency of the waveform that interacts with the component. So an inductor, for example, offers higher impedance as (a) its size increases and as (b) frequency goes up. (We will discuss the specifics of this in Chapters 5 and 6.)

What is not often stated, but is just generally understood, is that *frequency* in this context refers to a specific sinusoidal waveform of a single frequency, not something more complex. So when we say the impedance of a circuit is such-and-such a value at 1 MHz, we mean at a 1-MHz sine wave, not, for example, a 1-MHz square wave. That, then, means that when we analyze the performance of a circuit (using the basic formulas) we can do so only at a specific frequency and for a specific waveform (sinusoidal). This raises an interesting question: How do engineers analyze circuits when there are more complex waveforms?

The real answer, of course, is beyond the scope of this book. But in this part of the book I outline the basic approach that is used. Then, when you understand that, you will begin to understand some of the other subtleties we will talk about when discussing high-speed design issues.

There is a very important theorem in electronics called Fourier's Theorem. It forms the basis for many higher order analyses. It goes like this:

> **Fourier's Theorem:** Every signal or curve, no matter what its nature may be, or in what way it was originally obtained, can be *exactly reproduced* by superposing a sufficient number of individual sine (and or cosine) waveforms of different frequencies (harmonics) and different phase shifts.

This means that any waveform (in theory it must be a repetitive waveform, but in practice we often wink at that requirement) can be reproduced by superimposing a sufficient number of sinusoidal waveforms at higher frequency harmonics. So, we take a look at our input waveform of interest (perhaps a video signal, perhaps a complex digital signal formed by a series of square waves and pulses) and reproduce it with a series of sinusoidal waveforms (which Fourier's Theorem says we can do). Then we analyze the performance characteristics of our circuit for *each one* of the individual sinusoidal waveforms. Finally, we add up (superimpose) all of these results to get the combined result of the circuit to our input waveform.

Conceptually, this is simple and straightforward. From a practical standpoint, this is unmanageable, particularly since some waveforms (such as a square wave) require an infinitely long series of harmonically related terms. The way it is actually done involves very complex calculus and higher order transforms. But we don't have to deal with the complexities of how it is done. We simply need to understand that it can be done. Then we can appreciate the implications.

Let's look at some examples. A square wave is not a natural waveform that exists in nature. We can use a Fourier transform, however, to represent it as an infinite series of cosine or sine waves. One form of a square-wave series is as follows:

$$Square(\theta) = Cos(\theta) - \frac{Cos(3\theta)}{3} + \frac{Cos(5\theta)}{5} - \frac{Cos(7\theta)}{7} + etc \qquad [1\text{-}3]$$

Each component (term) in the series shown in Equation 1-3 represents a harmonic of the fundamental frequency, Θ. As it happens, a square wave, when represented by this cosine series, only contains odd harmonics. Figure 1-12 illustrates what happens as we use progressively more terms (harmonics) to represent the function. Each additional harmonic makes the resulting waveform look a little more like a square wave.

Figure 1-13 shows an example of this same series carried out to 101 harmonics (2 x 50 + 1). The example was created using Mathcad, a readily available software package available from Mathsoft, Inc. If you have Mathcad software, you can replicate this example on your own.

There is a demonstration program available on the UltraCAD Web site called Square.exe. Instructions for using it are included in Appendix A. It allows you to explore this cosine series in some detail and see how various assumptions affect the nature of the waveform.

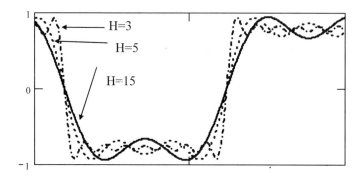

Figure 1-12 As we add more harmonics to the series, a better approximation to a square wave results.

Square Wave:
$$y(wt) := \sum_{n=0}^{50} (-1)^n \cdot \frac{\cos[(2 \cdot n + 1) \cdot wt]}{(2 \cdot n + 1)}$$

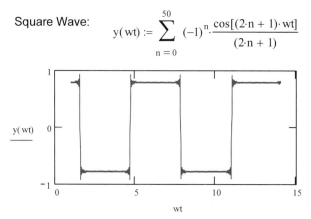

Figure 1-13 Fourier series for a square wave carried out in Mathcad.

Figure 1-14 illustrates various series for forming some other standard waveforms. What is instructive here is that any waveform can be "decomposed" into a series of sinusoidal harmonic terms that can then (at least conceptually) be individually analyzed, then summed, to determine the response of a circuit. Again, conceptually this is straightforward. In practice it can be quite difficult and involves very complex math and calculus.

If you understand this so far, then you can appreciate that there is a very significant implication of this for board design. Digital logic circuits deal with signals that look a lot like square waves. If we put a square wave onto a trace at one end, we want to get a square wave at the other end. That means our boards must be designed to handle not only the fundamental frequency of our digital signals, but also all of the higher order harmonics that are contained within it. Figure 1-12 suggests we want to be able to handle at least the tenth or fifteenth harmonics. So if we are dealing with a 10-MHz clock square wave, we need to be able to handle frequency harmonics up to at least 100 or 150 MHz. This is why we say it is not the (clock) frequency that matters, it is the rise time of the waveform and the harmonics needed to reproduce it that are important.

The term *bandwidth* is used to describe this requirement. The bandwidth is the range of frequencies we need to be able to deal with to faithfully maintain the integrity of our signals. One way to look at bandwidth requirements is suggested in the previous paragraph. Another way is suggested by Figure 1-9. Determine the fastest rise time you will encounter on the board and then determine the bandwidth requirement from the formula in Equation 1-4:

$$\text{bandwidth (Hertz)} \cong .3/\text{rise time} \qquad [1\text{-}4]$$

Sawtooth Wave $\quad y(wt) := \sum_{n=1}^{50} \frac{\sin(n \cdot wt)}{n}$

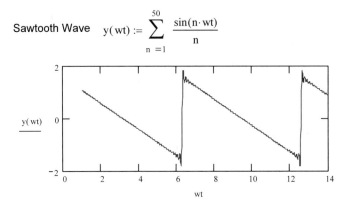

Triangular Wave: $\quad y(wt) := \sum_{n=0}^{5} \frac{\cos[(2 \cdot n + 1)(wt)]}{(2 \cdot n + 1)^2}$

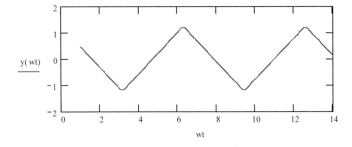

Impulse Signal:

$$y(wt) := 1 + \sum_{n=1}^{10} \cos(n \cdot wt) \quad y1(wt) := 1 + \sum_{n=1}^{30} \cos(n \cdot wt)$$

Figure 1-14 Fourier series for three common complex waveforms.

Chapter Endnotes

DIRECTION OF CURRENT FLOW

There is a point that I will mention once, here, and then we won't mention it again. I believe it was Benjamin Franklin who stated that current flows from positive to negative. And we have followed that convention ever since. But current is the flow of electrons, and electrons are negatively charged little critters that flow toward positive charges (or, perhaps more accurately, toward the absence of negatively charged particles). So electrons (current) flow from negative to positive, not from positive to negative! This anomaly has been with us since the beginning of electronic study and is usually ignored. We always use the convention that current flows from positive to negative, and the only time in my career when this has truly been an issue is when I have studied semiconductor theory, where the behavior of electrons is really at the heart of what is happening. So, if it ever does confuse you that there seems to be a contradiction here, well, Benjamin Franklin made at least one mistake in his life, and that has stuck with us for all time since!

SOME STANDARD SYMBOLS USED IN ELECTRONICS

Prefix	Symbol	Exponent	Multiplication Factor
exa	E	18	1,000,000,000,000,000,000
peta	P	15	1,000,000,000,000,000
tera	T	12	1,000,000,000,000
giga	G	9	1,000,000,000
mega	M	6	1,000,000
kilo	k	3	1,000
hecto	h	2	100
deca	da	1	10
(unit)		0	1
deci	da	-1	0.1
centi	c	-2	0.01
milli	m	-3	0.001
micro	u	-6	0.000001
nano	n	-9	0.000000001
pico	p	-12	0.000000000001
femto	f	-15	0.000000000000001
atto	a	-18	0.000000000000000001

SOME FUNDAMENTAL DEFINITIONS

They start with the electron or the flow of electrons:

1 amp = flow of 1 coulomb/sec
        where 1 coulomb = 6.25×10^{18} electrons

1 volt = (Electromotive) force (EMF) between two points for which a flow of 1 amp will generate 1 joule of energy (work)

1 watt = rate (power) at which work is done (joules/second)
      = volt x amp   (Note: 1 horsepower = 746 watts)

work = joule = power x time = watt x time
      (At home we pay for work in kilowatt-hours)

1 ohm = resistance that allows 1 amp to flow at 1-volt EMF

1 farad = storage of 1 coulomb of charge that will generate 1-volt EMF

1 henry = inductance that will cause an induced voltage of 1-volt EMF when the inducing current is changing at 1 amp/second

IMPORTANT NAMES IN ELECTRONICS

Name	Life Span	Symbol	Attribute
Charles Augustin de Coulomb	1736 – 1806	C	Charge
Andre Marie Ampere	1775 – 1836	A	Current
Alessandro Volta	1745 – 1827	V	Potential
James Prescott Joule	1818 – 1889	J	Energy, work
James Watt	1736 – 1819	W	Power
George Simon Ohm	1787 – 1854	W	Resistance
Michael Faraday	1791 – 1867	F	Capacitance
Joseph Henry	1797 – 1878	H	Inductance
Nikola Tesla	1857 – 1943	T	Magnetic flux density
Hans Christian Oersted	1777 – 1851	Oe	Magnetic field intensity
Wilhelm Eduard Weber	1804 – 1865	Wb	Magnetic flux

Others:

Carl Friedrich Gauss	1777 – 1855	
Gustav Robert Kirchhoff	1824 – 1887	
Heinrich Friedrich Emil Lenz	1804 – 1865	
James Clerk Maxwell	1831 – 1879	
E. L. Norton	1926*	
M. Leon Thevenin	1857 – 1926	

* Date Norton published his theorem

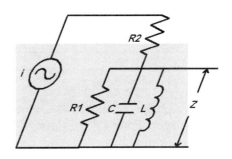

2 PROPAGATION TIMES

PROPAGATION SPEED

*E*lectrical signals traveling along a wire or trace in the air travel at the speed of light, 186,280 miles per second (which, if you do the arithmetic, works out to 11.8 in/ns). That speed is equivalent to 7.5 trips around the Earth's equator in one second. Yet we need to worry about how fast these signals travel along traces on our boards! That is pretty amazing when you think about it.

Consider this experiment: String a copper wire in the air across a lake and measure the time it takes a signal to travel from one end of the wire to the other end. Now lower the wire into the lake and measure the signal propagation time again. What do you think happens to the propagation time?

When a current flows along a wire or trace, it generates an electromagnetic field around the conductor. There are two components to this electromagnetic field, an electrical field and a magnetic field. The electrical field originates with the charged electrons. It radiates radially out from the conductor with a strength that decreases with the square of the distance from the conductor. The magnetic field originates from the motion of the electrons along the conductor. There is no magnetic field if there is no motion (current), and the strength of the field changes with the magnitude of the current. The magnetic field radiates around the conductor, with a polarity that depends on the direction of the current, and with a strength that is inversely related to the square of the distance from the conductor.

The electromagnetic field and the current must move together. The current can't move out in front of the fields. The magnetic field can't move out in front of the electric field. It turns out the issue is not how fast the current (electrons) can flow through the conductor. The issue is how fast the electromagnetic field can flow through the medium it flows through, the medium *around* the wire.

All materials have a characteristic called the *relative dielectric coefficient* (ε_r). It is a measure of the material's ability to store charge. It is called relative because it is referenced to the dielectric coefficient of air (really that of a vacuum, but the difference between air and a vacuum is so small that only astronomers care). Its relevance here is that the speed with which an electromagnetic field can travel through a material is equal to the speed of light divided by the square root of the dielectric coefficient of the material. Therefore, the propagation speed (in in/ns) of a signal is given by the formula in Equation 2-1:

$$\text{Propagation Speed} = 11.8/\sqrt{\varepsilon_r} \text{ in/ns} \quad [2\text{-}1]$$

So, back to our illustration of the wire dropped into the lake. The issue is not how fast the electrons can travel through the copper wire, the issue is how fast the electromagnetic field can travel through the water. The relative dielectric coefficient of water is approximately 80 (depending on its purity). The square root of 80 is approximately 9. So the signal will slow down when the wire is lowered into the water by about a factor of 9.

PROPAGATION TIMES

It is easy to confuse propagation *speed* and propagation *time*. Propagation speed is velocity. Its units are distance per unit time. Propagation time is, well, time. A related measure, wavelength (the distance a signal travels in one complete cycle), will be discussed later in this chapter. We express propagation time either in units of time (e.g., nanoseconds), or as units of time per unit length (e.g., nanoseconds/inch). Propagation time (expressed as time per unit length) is the inverse of propagation speed. That is:

Propagation Time (per unit length) = 1/Propagation Speed, or
Propagation Time = Length/Propagation Speed

For example, if an electrical signal in a copper wire in air travels at the speed of 11.8 in/ns, then it propagates down the wire at 1/11.8 or .085 ns/in.

Chapter 2

TRACE CONFIGURATIONS AND SIGNAL PROPAGATION

Figure 2-1 illustrates various trace configurations designers usually work with when designing high-speed boards. Most boards in situations where signal integrity is an issue have internal planes (as will be seen in almost every chapter in Part 2, "Signal Integrity"). From a signal propagation standpoint, traces that are located between two reference planes are considered to be in a stripline environment, regardless of whether the actual environment is a simple, centered, dual, offset, or asymmetric stripline configuration. A trace with a reference plane on only one side is considered to be in a microstrip environment.

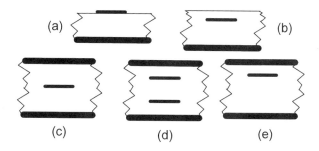

Figure 2-1 Common trace configurations on PCBs: microstrip (a), embedded microstrip (b), stripline (c), dual stripline (d), and asymmetric stripline (e).

A microstrip trace might simply be an external trace with air above it and board material between it and an underlying plane. An embedded microstrip trace has board material above the trace, also. This would particularly be the case if there were two external trace layers above the plane. At least the traces on layer 2 would be embedded microstrip traces. A coated microstrip trace would be one with board material between it and the reference plane, and with an additional material coat above the trace. The additional coat could be a variety of materials, including solder mask, conformal coating material protection, and so on.

We generally consider the material surrounding a trace in a stripline environment to be homogeneous (uniform). In fact, we usually specify it to be this way.

Therefore, the propagation speed of a signal in a stripline environment is dependably expected to be as described in Equation 2-1:

$$\text{Propagation Speed} = 11.8/\sqrt{\varepsilon_r} \text{ in/ns}$$

Actual experience generally only varies from this calculation if the ε_r is not as expected or if the material is quite nonuniform. Since the relative dielectric material of FR4 is approximately 4, and since the square root of 4 is 2, the propagation speed of a signal in a stripline environment is often considered to be $11.8/2 \cong 6$ in/ns, a rule we are all pretty familiar with. If we want to be more precise in our calculations, then we have to know the relative dielectric coefficient of the material we are using more precisely and plug that value into Equation 2-1.

The microstrip environment is more problematic. The material surrounding the trace is not uniform. In the simplest case it is air above the trace and dielectric below the trace. In more complex cases the dividing line between the dielectric and the air might not be precisely uniform, and there may be more than one type of material involved.

Therefore, if you want to estimate the signal propagation speed in microstrip you need to estimate the effective dielectric coefficient of the material surrounding the trace. There is a generally accepted rule of thumb for estimating the effective dielectric coefficient that has been in use for over 30 years.[1] That estimate is given in Equation 2-2:

$$\varepsilon'_r = 0.475\varepsilon_r + .67 \qquad [2\text{-}2]$$

There are a couple of problems with this estimate. The most significant one is that it is a constant. People have discovered that propagation time in microstrip is a variable, and, all other things being equal, is a function of trace width and height above the plane.

As trace widths get wider, propagation speed slows down! The reason for this is that as traces widen, a larger percentage of the field lines between the trace and the plane is contained within the dielectric. In the limit, with infinite trace

---

[1] There are numerous places where this equation can be found. See for example, IPC-D-317A, *Design Guidelines for Electronic Packaging Utilizing High-Speed Techniques*, p. 18. The reference given for this formula is H.R. Kaupp, "Characteristics of Microstrip Transmission Lines," *IEEE Trans.*, Vol EC-16, No. 2, April 1967.

width, virtually all the electromagnetic field is contained within the dielectric. In this case, the microstrip trace really looks very much like a stripline trace.

Microstrip propagation speed also slows down as the trace gets closer to the plane, for the same reason. A greater percentage of the field lines are contained within the dielectric than in the air. Brooks has shown that a much better estimate of propagation time in microstrip can be obtained by expressing propagation time in microstrip as a ratio of the propagation time for a signal in a stripline environment surrounded with the same dielectric material. He generates a formula for estimating this ratio.[2] The propagation *time* for a microstrip trace will never be longer than for a trace in a stripline environment surrounded by the same material. It will be shorter, depending on how the electromagnetic field splits between the air above the trace and the dielectric under the trace. His estimate for propagation time (in ns/in) is as follows:

$$\text{Propagation Time(Microstrip)} = Br \times \text{Propagation Time(Stripline)}$$

or,

$$\text{Propagation Time} = Br\sqrt{\varepsilon_r}/11.8 \qquad [2\text{-}3]$$

where:

$Br = .8566 + (.0294)Ln(W) - (.00239)H - (.0101)\varepsilon_r$
W = trace width (mils)
H = distance between the trace and the plane (mils)
ε_r = relative dielectric material between the trace and the underlying plane
Ln represents the natural logarithm, base e.

The fraction, Br, can never be greater than 1.0. A microstrip trace can never be slower than a stripline trace (surrounded by the same material). The formula was derived for simple microstrip traces with air above the trace and dielectric material underneath. Embedded and coated microstrip traces would have slightly higher fractions (and slightly slower propagation times) than estimated here, subject to an upper limit of 1.0.

---

[2] Brooks, Douglas G., "Microstrip Propagation Times: Slower Than We Think." Available for download at www.mentor.com/pcb/tech_papers.cfm.

Circuit Timing Issues

"Timing is everything" as the saying goes. In complex circuits we have signals on buses running everywhere. There are certain moments in time when those signals must align exactly right. Just a few of those times might include:

1. The three colors of a video display (red, green, and blue) must arrive at a receiver at the same time or the video picture will be distorted.
2. The two traces of a differential pair must be the same length so that the two opposite signals arrive at the receiver at the same time or common mode noise may be generated (see Chapter 14).
3. The data lines must be present and stable at their respective gates when the clock pulse triggers or there will be data errors.

Figure 2-2a illustrates what might happen if there is a timing problem. Three signal lines are shown, the top slightly misaligned from the others. A clock pulse samples the data state. The data signals don't line up quite right, but the clock signal still samples the data signals when they are all high or all low. In Figure 2-2b, however, the top signal is so badly misaligned that it transitions during the time the clock is sampling the data lines. In Figure 2-2b there will be data logic errors.

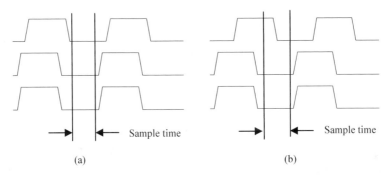

Figure 2-2 Circuits can tolerate slight differences in timing (a), but if the differences in timing are too great, sampling errors will occur (b).

Signals get misaligned for several reasons. One of them involves the devices themselves. There is a throughput time for a signal to travel through a device.

There are tolerances on output timing of devices, and circuits have different numbers of devices the signals propagate through. There are propagation delays on traces themselves. Yet, at certain well-defined points and times in a circuit and system, all the signals must align properly. The *specification* for this is the responsibility of the circuit design engineer. But *making it happen* can be the responsibility of the board designer.

Board designers control signal timing with trace length. Multiplying the propagation time (in units of ns/in) of a trace by its length gives the propagation time of the signal down the trace. If we need the trace to be a specific increment of time, then we adjust its length to make it so. This process is often called *tuning* the trace.

Adjusting timing this way requires two things, (a) a precise knowledge of the propagation speed of the signal along the trace we are interested in, and (b) the ability to adjust the trace length. We determine the propagation speed from the preceding discussion. Inherent in the determination is the knowledge of the relative dielectric coefficient (ε_r) of the material surrounding the trace. We cannot make the trace shorter than the distance between its connection points, but we can make it longer. The normal way to adjust trace length is to "snake" it.

Many of the higher end PCB design programs will automatically adjust trace length once you give it the appropriate parameters. Figure 2-3 illustrates the tuning editor from one of them. Note the variety of patterns the designer can specify. Figure 2-4 illustrates a portion of a board UltraCAD Design, Inc. has designed that required a considerable amount of tuning.

Some engineers worry about EMI radiation that might be caused by tuning patterns. In our experience we have seen no EMI issues from any type of tuning pattern as long as (a) every trace is referenced to a signal plane, and (b) the tuning is confined to stripline signal layers. Lots of traces are successfully tuned on microstrip layers (look at almost any computer motherboard), and we do so all the time. But if you or your engineers have significant concerns about tuning causing EMI problems, confining tuning to inner layers will almost always eliminate any possibility of radiation.

It should be noted that there are some articles that suggest crosstalk between segments of tuning loops can interact with the signal propagation speed and distort timing calculations. This is probably a concern at only the highest clock

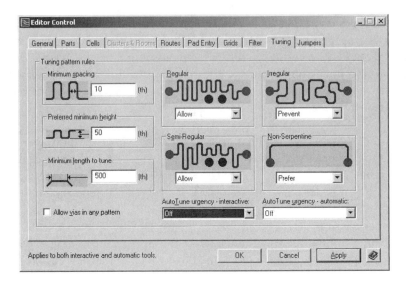

Figure 2-3 Tuning options from a PCB design software package.

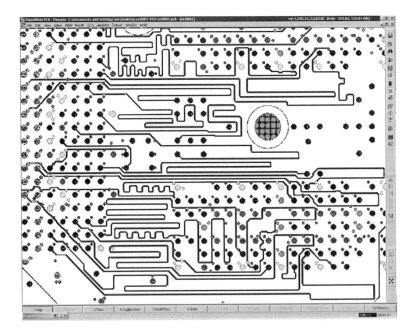

Figure 2-4 Tuned traces on a circuit board.

frequencies and is beyond the scope of this book. We have heard many people argue about whether or not via length should be taken into consideration while calculating propagation time. There does not seem to be a definitive conclusion in the industry about this.

WAVELENGTH

The wavelength of a signal, λ, is defined as the distance a signal travels during one complete cycle (see Figure 2-5). In air, this is simply the speed of light divided by the frequency in Hertz (Equation 2-4):

$$\lambda = c/f \qquad [2\text{-}4]$$

where λ = wavelength
c = speed of light in units of length per second
f = frequency in Hertz

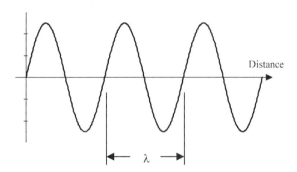

Figure 2-5 Wavelength, λ, is the distance the signal travels in one complete cycle.

On circuit boards, we need to know the propagation speed of the signal on the board before we can calculate wavelength. Then we would substitute that value for c in Equation 2-4. For example, the wavelength of a 300-MHz (which is .3 cycles/ns) signal in an FR4 stripline environment ($\varepsilon_r = 4$) would be:

From Equation 2-1:

$$\text{Propagation speed} = 11.8/\sqrt{4} = 5.9 \text{ in/ns}$$

Then from Equation 2-4:

$$\lambda = 5.9/(.3) = 19.7 \text{ in.}$$

Wavelength is normally not an important parameter in board design. But it arises in at least two types of discussions. The first is if we are designing a transmitting antenna onto the board. (Note: Receiving antennas are not nearly as dependent on wavelength as transmitting antennas are.) This might be the case for a cell phone, pager, automotive door locking device, and so on. Transmitting antennas tend to perform best when they are some (fractional) multiple of a wavelength. Therefore, we need to know propagation speed in order to calculate wavelength and therefore an optimum length for the antenna we are putting on the board.

Note that if we are designing an antenna onto the board and are concerned about wavelength, then it is important that we *specify* the relative dielectric coefficient of the board material to be used in the fabrication of the board. In the case of microstrip traces we must also specify the type and extent of any coatings that might be used over the traces. Since propagation speed is dependent on relative dielectric coefficient, we cannot be certain what the speed will be, and therefore the wavelength of the signal, if we do not control the material. We control the material through a material specification that is part of the fabrication drawing for the board. More than one company has fallen into the trap of not controlling the material specification and then wondering why the effective transmitting length or performance of their boards changes from production lot to production lot.

Another context in which board designers may see wavelength mentioned relates to via spacing. In Chapter 9 we mention the concept of a Faraday shield in the context of controlling EMI emissions from an inner layer of a board. Some people advocate placing a row of vias along the edge of the board to trap EMI emissions within the board. If so, then they usually recommend that the vias be spaced some fraction of a wavelength (often 1/20).

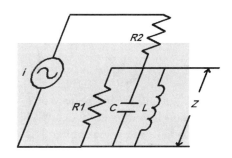

3 ELECTRICAL COMPONENTS

THE THREE BASICS

*W*hen we think about all the challenging complexities of high-speed designs, it is sometimes hard to believe how few components there really are in the world of electronics. Are you ready for this? There are only three passive components we will ever have to worry about in high-speed design issues. Only three! They are (a) resistors, (b) capacitors, and (c) inductors. Virtually everything we encounter in board designs can be described and analyzed in terms of these three components and their combination and effects.

To go with them, we may need an energy source. If so, we can consider it to be either a pure current source or a pure voltage source (an example of a voltage source might be a battery). And, as we will see later, the fundamental nature of inductance is electromagnetic energy, which relates to EMI and crosstalk. But *pure* energy sources are simple concepts, and electromagnetic energy is at the heart of inductance. So this does not negate the statement that we need understand only three basic passive components.

We usually consider semiconductor devices to be active components. But even these, in steady state, can be analyzed in terms of a simple energy source and just the three passive components, as we shall see. A clock driver or a logic gate at its "one" or "zero" output level, for example, can be considered to be at a steady state. (Note: A semiconductor device during switching between states, an amplifier with varying input, and circuits with feedback—such as control loops—often, but not always, are a little more difficult to analyze.)

So why are high-speed design issues so difficult? Well, actually, they aren't so difficult if you really understand the behavior of the three basic components and how they interact. The problem is that many people, even many degreed engineers, do not really understand them. And that is a shame, because that single, simple truth is in large part the reason for so much confusion and mystery about high-speed design problems.

This chapter focuses on the three basic components and their fundamental natures. Hydraulic analogs are used as examples. The next chapters will cover their electrical effects (resistance and reactance) and what happens when we combine resistance and reactance to get impedance.

The discussions in this chapter relate to "ideal" components: ideal resistors, capacitors, and inductors. Ideal components are "pure" components. That is, an ideal resistor is a pure resistor with no (parasitic) capacitance or inductance associated with it. An ideal inductor has no resistance or capacitance associated with it, and so on.

In reality, of course, no component is ideal or pure. But once we learn to deal with these ideal components, and their combinations, we will then know how to deal with more realistic situations (e.g., capacitors with lead inductance).

RESISTANCE

Remember, current is the flow of electrons. Electrons (typically) flow through a wire or trace. A hydraulic analog would be water flowing though a pipe. In order to "push" the electrons along a wire we need an energy source. Figure 3-1a shows a battery as the energy source. Similarly, water needs a force to push it through the pipe. Figure 3-1b shows a pump as the source of the force pushing water through a pipe.

A resistor affects the flow of the electrons. (Of course, so do capacitors and inductors.) Its effect is to impede (related to impedance) the flow. The analog would be a crimp in the pipe that impedes the flow of water. A small crimp (resistance) offers a little impedance to the flow, whereas a bigger crimp offers a lot of impedance. A very large crimp (resistance) might almost stop the flow altogether.

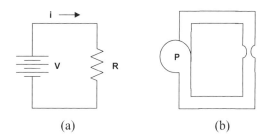

Figure 3-1 Resistance can be modeled by a crimp in a hydraulic pipe.

A small crimp might allow a certain flow of water. A larger crimp would reduce the flow and cause pressure to build up behind the crimp, effectively fighting the pump pressure. If we wanted to maintain the same flow with the tighter crimp (bigger resistance) we would have to provide more pump force (voltage) to overcome the pressure backing up behind the crimp. Thus, there is a relationship between the size of the crimp (resistance), the flow of the water (current), and the pump force (voltage).

OHM'S LAW

There is an extremely important formula that relates the flow of electrons through a resistor. It is called Ohm's Law (after George Simon Ohm):

$$V = i \times R \quad\quad [3\text{-}1]$$

where V is the voltage force (in volts)
i is the current (in amps), and
R is the resistance (in ohms, or Ω)

Other variations on Ohm's Law are:

$$i = V/R$$
$$R = V/i$$

Given any two values, the third can be calculated from Ohm's Law. For example, if V = 5 volts and R = 1K (1,000 Ω), then i will be 5 mA (.005 amps).

This is perhaps the most important formula in the book! If there is only one thing in this book you need to commit to memory, this is it. Absolutely anyone and everyone in our industry needs to be able to write down and work with Ohm's Law without having to think about it.

There are a couple of simple observations we can make by looking at Figure 3-1. For example, the flow of water (and the current) must be the same at every point in the pipe (or wire or trace). If this were not true, where would the electrons or the water go? You can't pump water out of a pump if you don't have water coming into the pump. And you can't force electrons out of a battery if they are not also flowing in.

Consider the potential impact if this were not true. If we could force electrons out of the battery without returning them in the other side, then somewhere in the circuit there would be a big buildup of electrons and somewhere else there would be a big shortage. These two areas would then have large negative and positive charges, respectively. Then the two areas would really be hard to handle, attracting or repelling lots of other things around them. We know intuitively that this doesn't happen. So, current flows in a loop, and when it does so it must be the same everywhere in that loop.

We will see later that current might divide into branches, and we will discuss the laws relating to that at that time. But current in any branch is the same at every point in the branch, and ultimately returns to its source.

A second observation relates to the forces around the loop. Consider the voltage at the battery in Figure 3-1 (or the pressure at the pump). It is the same as the voltage across the resistor (or the pressure across the crimp). If we were to go around the loop and add up the voltages (or pressures) they would sum to zero. For example, if we went clockwise around the loop the voltage across the resistor would be +V and the voltage across the battery would be measured as –V (in a clockwise rotation we would see the negative terminal of the battery first). This is a simple illustration, but nevertheless it is a fundamental truth that voltages around even a more complex loop must always sum to zero. Always. This is known as Kirchhoff's (Gustav Robert Kirchhoff) 2nd Law (see Chapter 5).

Let's now look at two resistors (crimps) in series (Figure 3-2). If one resistor (crimp) is very large, very little will flow, regardless of how small the other resistor is. For there to be a large flow, both resistors (crimps) need to be relatively small. We will see later that the effect of two resistors (crimps) in series is additive; that is, the equivalent resistance, Req, is given by Equation 3-2.

$$Req = R1 + R2 \qquad [3\text{-}2]$$

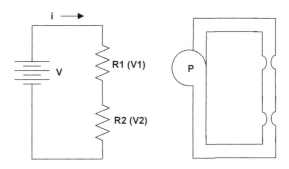

Figure 3-2 Hydraulic analog of two resistors connected in series.

Of particular note, if R2, for example, is very much larger than R1, then the equation for equivalent series resistance reduces to Equation 3-3.

$$Req = R1 + R2 \cong R2 \qquad [3\text{-}3]$$

If we put two resistors (crimps) in parallel, as in Figure 3-3, the situation is a little different. First, notice that the current out of the battery must equal the current into the battery, as we discussed before, just as the water flowing out of the pump must equal the water flowing into the pump. But the current (water) divides into two branches. Since we can't create current (or water) from nothing, the sum of the current (water) in the two branches must equal the current (water) through the battery (pump).

In these figures, there are two nodes: one at the top where the two branches break off from the main line, and the other at the bottom. The current (water) into a node must equal the current (water) out of the node. If this were not true, where would the electrons (or water) come from or go?

Also, since there can only be a single voltage (pressure) at a node, then the voltage across R1 must be the same as the voltage across R2.

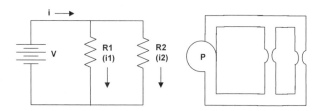

Figure 3-3 Hydraulic analog of two resistors connected in parallel.

Now consider what happens if R1 and R2 are significantly different, say R1 is very much larger than R2. In that case, most of the current will flow through R2. In fact, no matter how large R1 is, the effective resistance of their combination will never be greater than R2. The equivalent resistance of their combination will be smaller than R2 as R1 reduces in size. It is reasonably intuitive to see that if R1 = R2, then the current through R1 will equal the current through R2 (since the voltage across them is the same, and in fact equals V in this illustration.) If i1 = i2, then the equivalent (single) resistor that looks like the parallel pair of resistors R1 and R2 (where R1 = R2) would be R/2. The formula for the equivalent resistance, Req, of two parallel resistors is given by Equation 3-4:

$$\text{Req} = \frac{1}{\frac{1}{R1} + \frac{1}{R2}} = \frac{R1 R2}{R1 + R2} \quad [3\text{-}4]$$

This is called the *parallel combination* of two resistors. We will see later that a similar formula will apply to some combinations (not necessarily parallel ones) of inductance, capacitance, reactance, and impedance.

Of particular note, if R2 is very much larger than R1, then the equation for the equivalent parallel resistance reduces to Equation 3-5 (compare that with Equation 3-2).

$$\text{Req} = \cfrac{1}{\cfrac{1}{R1} + \cfrac{1}{R2}} = \cfrac{R1R2}{R1+R2} \cong R1 \qquad [3\text{-}5]$$

CAPACITANCE

Capacitors are interesting devices. The hydraulic analog to a capacitor is a storage device with two separate, nonconnected, sealed chambers (Figure 3-4). As a pump forces the water around the pipe, water flows into the top chamber and is pumped out of the lower chamber. At first, this happens with almost no impedance to the flow. But as the chamber begins to fill, the pressure in it builds up, pushing against the pressure of the pump and causing the flow of water to slow. When the chamber fills, the pressure at the chamber equals the pressure at the pump and the flow stops.

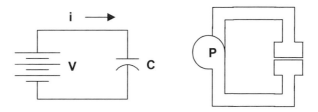

Figure 3-4 Hydraulic model of a capacitor.

A capacitor, in its simplest form, consists of two parallel plates placed closely together. Actual fabrication techniques are, of course, more complicated than this. But all capacitors, fundamentally, consist of two separate surfaces placed very close together with a thin material placed between them.

When a voltage is applied to a capacitor (perhaps by connecting a battery across its leads), electrons flow onto one of these plates. (Remember, the flow of electrons is current.) Electrons flow off the other plate and back around into the battery. This preserves the condition that current is the same everywhere in the loop. During the first instant, the flow is virtually unimpeded. But as the electrons begin to build up on the plate, they start building up a charge that begins to resist the continuing flow of more electrons. Thus the flow starts slowing down. At some point, the charge buildup is great enough to stop any further flow, and all current stops. The charge buildup is manifested by a voltage on the plate. The voltage increases as more charge builds, and current stops flowing when the voltage equals the initial voltage that was applied to the capacitor.

One key to understanding what happens with a capacitor is the recognition that what happens is *time dependent*. (This is *not* true with resistance.) That is, when current first starts flowing into a capacitor, there is no voltage across the plates (i.e., across the capacitor). There can be no voltage across the capacitor until charge builds up on the plates. But charge can't build up on the plates until charge (current) has actually flowed into the capacitor. Since current must flow before a voltage appears, we say that current leads voltage. (Why this concept is important will become more apparent when, later on, we see that the opposite is true with inductors; that is, current lags voltage in the case of an inductor.)

Now, consider the hydraulic analog and think about what happens if we turn the pump on in the forward direction for a moment and then reverse it, pumping in the opposite direction. Water will start to flow into the storage tank, but if we reverse the pump soon enough, negligible pressure will have built up. Then water will flow in the reverse direction. Now if we reverse the pump again, quickly enough, pressure will not build in the opposite direction either.

In this way, we can envision how it would be possible to have a pump that continually reverses direction fast enough that water pressure cannot build up in the tank. The flow of water in the pipes will *look* as though the storage tank were not there. Of course it is, but the flow reverses so quickly that its effects are not seen. We would see this same effect if the pump reversed more slowly, but the storage tank was larger. But if the pump reversal was relatively slow, and the storage tank was relatively small, then we would see the flow of water speed up and slow down in different parts of the pumping cycle as pressure built up and then decreased again.

So, the effect of the storage tank is a function of both the size of the tank and the frequency at which we alternate the pumping cycle. Water will flow as if the tank were not there if the tank size is very large or if the pump frequency is very high. Otherwise, water flow will slow down, and might even stop, during part of the cycle.

So it is with electrons and a capacitor. If a capacitor is very large, alternating current will flow virtually unimpeded, as if the capacitor were not there. If the frequency is very high, current will flow as if the capacitor were not there. It is not that *individual* electrons are flowing around the loop; they are not. Individual electrons cannot travel past the plate of the capacitor. But the flow of electrons past any particular point will be the same as it would be if the capacitor were not present. On the other hand, there will be significant impedance to current flow if (a) the frequency is relatively low, or (b) the capacitor is relatively small.

Another way to think about this is to recognize how many electrons actually flow during a half-cycle of an AC current. A 1-MHz waveform changes direction 10^6 times each second. Therefore, for the same magnitude of peak current (charge/unit time), only one-millionth (10^{-6}) as many electrons flow during the half-cycle of a 1-MHz (10^6) waveform as flow during the half-cycle of a 1-Hz waveform. Therefore, the voltage buildup on a capacitor will be substantially less for a half-waveform of a 1-MHz signal than for a 1-Hz signal. Table 3-1 summarizes the range of possibilities.

Frequency

		High (Fast)	Low (Slow)	DC
Cap	Big	Lots of flow	Some flow	No flow
	Small	Some flow	Very low flow	No flow

Table 3-1 Capacitor size and frequency both affect the flow of current.

Suppose we put two capacitors in parallel (Figure 3-5). If the value of one capacitor is very large, then it almost doesn't matter what the value of the other one is. Even if the other capacitor is very small, almost zero, the overall circuit will still behave as if there is a large capacitor in the flow loop. The flow will be determined, first, by the larger capacitor. The effective capacitance of the two will be at least as large as the larger of the two capacitors, and perhaps even more if the second capacitor is nearly the same size as the first.

In contrast, consider the case of two capacitors in series (Figure 3-6). Here, the flow will be restricted by the smaller capacitor. The smaller capacitor will charge up quickly (the smaller storage tank will fill up quickly) and flow will stop. No matter how large the larger capacitor is, the flow will never be greater than that allowed by the smaller capacitor. The total flow may be less if the other capacitor adds additional restriction (impedance to the flow), which would happen if it were nearly as small as the first one.

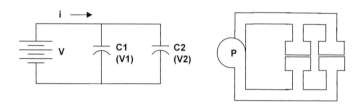

Figure 3-5 Hydraulic analog of two capacitors connected in parallel.

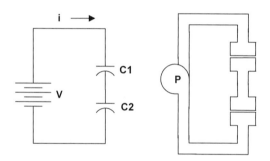

Figure 3-6 Hydraulic analog of two capacitors connected in series.

For the parallel combination of capacitors, Ceq is given by Equation 3-6:

$$Ceq = C1 + C2 \qquad [3\text{-}6]$$

Note that if, for example, C2 were very much larger than C1, the equivalent parallel combination would reduce to Equation 3-7:

$$Ceq = C1 + C2 \cong C2 \qquad [3\text{-}7]$$

For the series combination of capacitors, Ceq is given by Equation 3-8:

$$Ceq = \frac{1}{\frac{1}{C1} + \frac{1}{C2}} = \frac{C1 C2}{C1 + C2} \qquad [3\text{-}8]$$

Note in Equation 3-8 that if C2, for example, were very large, the equation would reduce to Equation 3-9

$$Ceq = \frac{1}{\frac{1}{C1} + \frac{1}{C2}} = \frac{C1 C2}{C1 + C2} \cong \frac{C1 C2}{C2} = C1 \qquad [3\text{-}9]$$

This illustrates how the smaller capacitor becomes the limiting one in this case. Note that this is the opposite case from that of resistors (Equations 3-3 and 3-5).

The measure of capacitance is the farad (Michael Faraday). One farad of capacitance will have a force of one volt across its plates when the capacitor stores one coulomb of charge. (Remember one coulomb is 6.25×10^{18} electrons.) Recall also that one amp is the flow of one coulomb of charge in one second. So, if one amp of current flows for one second onto the plates of a one-farad capacitor, the voltage across the plates will rise one volt, or:

$$\Delta V = i \times \Delta t / C$$

Perhaps of more relevance, what if 1.0 mA flows onto the plate of a 10.0 uF capacitor for 1.0 ms? Well, 1.0 mA (10^{-3}) for 1.0 ms (10^{-3}) would move 10^{-6} coulombs. A capacitance of 10.0 uF is 10^{-5} farads. So the voltage buildup would be:

$$V = \text{charge/capacitance}$$
$$= 10^{-6}/10^{-5} = 0.1 \text{ volts}$$

CHARGE STORAGE

Suppose we charged a capacitor up to 10 volts and then disconnected it from the circuit. How long would the capacitor hold the charge? Conceptually, a perfect capacitor would hold the charge forever. In practice the charge would decay through time, primarily because of what we call *leakage resistance*. Almost all capacitors will slowly leak charge internally from one plate to the other. The mechanism for how this happens is a function of the materials and the fabrication processes used. The speed with which this happens is determined by the magnitude of the leakage resistance, which is usually part of the specification for the capacitor or capacitor family. Leakage resistance is usually quite high and not an issue in high-speed designs except in very special cases.

If we had a charged, disconnected capacitor sitting on a workbench and wanted to see how fast it was discharging (how fast the charge was leaking off), this would be hard to measure. Most measurement tools (especially multimeters and scopes) absorb charge during measurement, therefore affecting the very effect you are trying to measure.

Certain metal-oxide-silicon (MOS) devices (e.g., some charge coupled devices) use little silicon capacitors as "charge buckets" to store digital information. Some devices can be designed to hold charge for a considerable time, allowing them to function as nonvolatile memories (i.e., memories that don't lose their information when the power supply is turned off).

A more serious consideration involves capacitors in high-voltage circuits normally associated with CRTs, laser printers, and copiers. These capacitors may be charged to many thousands of volts and may still hold this charge when the device is turned off. Good safety practice is to place a very high resistance "bleeder" resistor around such capacitors so that they will discharge safely within a few minutes. For various reasons, these resistors may be missing or ineffective in any particular circuit. Service personnel learn quickly (if they don't already know) to be very careful around such circuits when servicing them. An inadvertent shock from a highly charged capacitor is, at a minimum, uncomfortable, and can very easily be quite dangerous.

FORMULA FOR CAPACITANCE

The formula for capacitance is fairly straightforward (see Equation 3-10). It is purely a function of the geometry of the plates and the material between the plates. Some capacitors (e.g., ceramic ones) are made up of several layers of plates, and the n − 1 term accounts for this.

Chapter 3   49

$$C(pF) = .2248 \times \varepsilon_r \times A \times (n-1)/d \qquad [3\text{-}10]$$

where: ε_r = relative dielectric coefficient (air = 1, FR4 ≅ 4)
A = area of each plate (in^2)
d = distance between plates (in)
n = number of plates

Applying this formula to a PCB made from FR4 with ε_r = 4, 3-inches square, and with power and ground planes spaced 10 mils apart would lead to a capacitance of

$$C = .2248 \times 4 \times 9 \times 1/.01 = 809 \text{ pF}$$

CAPACITANCE FUNCTIONS AND EFFECTS

The capacitance between closely spaced power and ground planes on a good-sized PCB can be fairly significant. We will see later that this might be used to our advantage in high-speed designs. If the planes are spaced further apart, however, the available capacitance drops quickly. We cannot just blindly assume the capacitance formed by the planes will work to our advantage. Some conditions do apply.

Capacitors provide several different important functions in a circuit. Their ability to store DC charge and to offer relatively low impedance to alternating frequency signals make them useful in power supply circuits and bypass applications. We can think of them as functioning as little batteries here. Since they tend to block DC but allow AC to pass, they are used to "couple" circuit stages together. The signal can pass from one stage to the next, but DC bias voltages are isolated to their own stage. Since the amount of charge that accumulates on their storage plates is a function of time and current flow, they can be used to "time" various circuit functions, such as the length of time a digital camera shutter is open or a circuit gate is open to accept a signal. We will see later that capacitors can be combined in a unique way with inductors to control frequency. Thus they are an important part of tuning and filter circuits.

But sometimes capacitors show up where we don't want or expect them to. For example, consider two traces routed side by side. Their edges can form the plates of a capacitor, and a signal can be coupled from one trace to the other, creating unwanted noise on the other trace. This is an important cause of crosstalk. Wirewound resistors have so much capacitive coupling between their wire turns that they are virtually useless, and almost never used, in higher frequency circuits. Unwanted and undesirable capacitive effects are among the top three problems in high-speed PCB designs (the other two are inductance and EMI).

INDUCTANCE

Inductors seem to be very difficult components for many people to understand. Appendix B is devoted to explaining why inductors do what they do (induct). Here, I try to give an intuitive understanding of what happens when there is an inductor in a circuit.

The hydraulic analog for an inductor (Figure 3-7) is a paddle wheel with significant mass. When the pump applies pressure through the pipe, the paddle wheel's inertia resists the initial pressure, and pressure builds up behind it in the pipe. The pressure begins to overcome the inertia, and the paddle begins to turn. As long as there is pressure behind it, the paddle wheel will continue to speed up until it reaches an equilibrium with the flow.

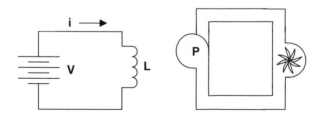

Figure 3-7 Hydraulic analog for an inductor.

Most of us probably don't see many paddle wheels, so let's consider a perhaps more familiar illustration of inertia first. When your car is sitting in the driveway and won't start, think of what happens when you try to push it. It is very hard to get started, and you have to push it very hard to get it started. But once it starts to roll, it can be equally hard to stop. Some of us have had the unfortunate experience of having our car roll into another car or into the side of our garage before we could stop it again.

I had an opportunity once to open and close a huge (5,000 lb.) bank vault door. It was beautifully balanced on its hinges, and swung easily after I got it started. It took considerable effort to stop it again before it swung so far that it closed. It was a beautiful example of almost pure inertia.

So now think what happens when we turn on the pump (voltage). Pressure builds instantly against the vanes of the paddle wheel (inductor) but no water (current) flows. The pressure must "work" against the inertia for a moment before the wheel begins to turn. After a period of time, the wheel begins to turn and allow water (current) to flow. How long this takes is a function of how much force is applied and how much inertia there is in the paddle wheel. The force will continue to speed up the wheel until the wheel reaches an equilibrium speed such that all the water will flow unimpeded around the loop and there will be no pressure at all across the wheel.

Now, suppose we turn off the pump. The paddle wheel's inertia will cause it to continue to turn, forcing water to continue to flow around the loop. It is as though the paddle wheel takes over the pump's function. Reverse pressure will build up across the pump as water is forced to its "back" side and drawn away from its front. This will continue until the force that is built up is finally able to counteract the inertia of the paddle wheel, and the paddle wheel slows to a stop.

So it is with electrons. When we place a battery across an inductor, there is force (voltage) across the inductor, but in the first instant there is no current flow. Current does begin to flow, however, and after a moment the current will reach an equilibrium and flow unimpeded around the loop. At that point the voltage across the inductor has dropped to zero. How quickly the current begins to flow and how quickly it reaches equilibrium depend on how large the inductor is. It takes longer for this to happen with a larger inductor than it does with a smaller inductor.

The impedance to the initial flow of current has its source in the buildup of the magnetic field around the conductor. For a more extended discussion of this characteristic see Appendix B, "Why Inductors Induct." After the current stabilizes there is a stable magnetic field around the wire. Now when we remove the source of current, the magnetic field around the conductor begins to collapse, generating the continuing force to keep current flowing in the same direction. Electrons will continue to flow and build up on the far side of the inductor. As they build up, the charge (voltage) increases. When the voltage increases enough, it will repel the further flow of electrons. The force will finally balance against the "inertia" of the inductor. At that point, the flow will stop.

So, as with a capacitor, what happens to the relationship between voltage and current with an inductor is time dependent. At the instant when voltage is first applied, voltage appears across the inductor but no current flows. Thus, voltage *leads* current, or current *lags* voltage through an inductor. In this sense, an inductor behaves exactly the opposite as a capacitor.

Also, in contrast to a capacitor, consider what happens when we pump water in one direction, then reverse the pump and pump it in the other direction. If we reverse the direction of the pump quickly, before the "inertia" can be overcome, very little water flows at all. On the other hand, if we pump for a longer time in one direction, then reverse the pump and pump in the other direction for a while, a reasonable flow will exist. A large inertia (inductor) will resist the flow more and for a longer time, and a smaller inertia will let the flow start more quickly.

Thus, the effect of the inductor is a function of both the size of the inductor and the frequency. Flow is small if the inductor is large or if the frequency is high; flow is larger if the inductor is smaller or the frequency is lower (see Table 3-2).

		High (Fast)	Low (Slow)	DC
L	Big	Very low flow	Some flow	Full flow
	Small	Some flow	Lots of flow	Full flow

Table 3-2 Inductor size and frequency both affect the flow of current.

Suppose two inductors are connected in series (Figure 3-8). If L1 is much larger than L2 (i.e., the first paddle wheel is much larger than the other), the current (flow) cannot be larger than that allowed by the larger one. The second, smaller inductor will not have much effect, unless it is about the same size as the larger one.

By the time the force can overcome the inertia of the larger one, the smaller one's inertia has already been overcome.

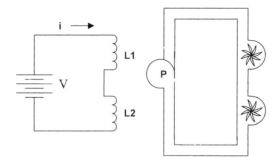

Figure 3-8 Hydraulic analog of two inductors connected in series.

As is the case with resistors, inductors in series add together. So the equivalent inductance, Leq, if two inductors are connected in series is given by Equation 3-11:

$$\text{Leq} = L1 + L2 \qquad [3\text{-}11]$$

Note that if L2 is very much larger than L1, then the equation for Leq reduces to Equation 3-12:

$$\text{Leq} = L1 + L2 \cong L2 \qquad [3\text{-}12]$$

If there are two inductors in parallel, however (Figure 3-9), the current will not be less than what the smaller one will allow. The smaller inductor's inertia is quickly overcome by the force, so current flows through it whether or not the inertia of the larger one has been overcome. If the other inductor is small enough, additional current can flow through it. Thus, parallel inductors combine into a single equivalent, Leq, in the same manner that parallel resistors do (see Equation 3-13):

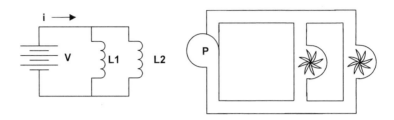

Figure 3-9 Hydraulic analog of two inductors connected in parallel.

$$\text{Leq} = \frac{1}{\frac{1}{L1} + \frac{1}{L2}} = \frac{L1 L2}{L1 + L2} \qquad [3\text{-}13]$$

Note also that if one inductor, say L2, is very large, the equivalent parallel inductance simply reduces to Equation 3-14:

$$\text{Leq} = \frac{1}{\frac{1}{L1} + \frac{1}{L2}} = \frac{L1 L2}{L1 + L2} \cong \frac{L1 L2}{L2} = L1 \qquad [3\text{-}14]$$

Note that inductors combine exactly as resistors do. Inductors and capacitors are opposites in this characteristic also.

The measure of inductance is the henry (Joseph Henry). Remember the measure of capacitance is the farad, and the voltage across a capacitor will rise one volt when one amp flows onto the plates. A current flowing through an inductor may generate a voltage across it, also. But from the preceding it is clear that, in the case of an inductor, we are not talking about a DC current. A DC current does *not* generate voltage across an inductor just as a steady flow past the paddle wheel does *not* need a force to keep it going (once the paddle wheel is up to speed). Force builds up across an inductance when the current flow through it is changing, working against the inertia of the inductance. One henry of inductance will cause one volt to appear across it when the current through it is *changing* at the rate of one amp per second, or, more generally:

Chapter 3

$$V = L \times \Delta i / \Delta t$$

FORMULAS FOR INDUCTANCE

Formulas for inductance are moderately complicated and are also somewhat approximate. Equations 3-15 and 3-16 give generally accepted formulas for a wire in air and for a microstrip trace:

Wire in air:

$$L = .00508\,b \left\{ \left[\mathrm{Ln}\left(\frac{2b}{a}\right) \right] - .75 \right\} \qquad [3\text{-}15]$$

where   L = inductance in µH
        a = radius (in)
        b = length (in)
        Ln = natural log (base e)

For example, let's assume a typical through hole wire is .018 in in diameter. Then a 1.0 in length would have an inductance of:

$$L = .00508 \times 1 \times \left\{ \left[\mathrm{Ln}\left(\frac{2 \times 1}{.009}\right) \right] - .75 \right\} = .012\ \mu H = 12\ nH/in$$

Microstrip trace:

$$L \cong 5\,\mathrm{Ln}(2\pi h/w) \qquad [3\text{-}16]$$

Where: L = inductance in nH/in
        w = trace width (mils)
        h = height above the plane (mils)

For example, a 10-mil-wide trace 1.0 in long and 10 mils above the plane would have an approximate inductance of

$$L \cong 5\,\mathrm{Ln}(2\pi 10/10) \cong 9.2\ nH/in$$

Note that these component lead and trace inductances are pretty small, on the order of 9 to 12 nH per inch. But if the current is rising at, say 10 mA per ns, (which might be representative of a 1-ns rise time device) the voltages generated can be surprisingly large:

$$V = L\Delta i/\Delta t$$
$$V = (12)(10^{-9})(10^{-2}/10^{-9}) = .12 \text{ volts} = 120 \text{ mV}$$

This gives an idea of why stray inductance is a signal integrity problem in high-speed designs. Ordinarily 9 to 12 nH is an inductance that would be considered negligible (unless we were talking about RF or microwave circuits). But when we talk about nanosecond rise times, such inductances can be a problem, indeed.

High-speed digital circuits don't usually have very many inductors in them, but inductors are an integral part of electromechanical relays and transformers. They may be used in power supply filtering circuits because of their ability to pass DC power supply voltages while "blocking" (thereby filtering out) higher frequency noise. Sometimes a simple ferrite bead is used to provide localized inductance for this purpose. Inductors sometimes are used to set frequencies or as part of filtering circuits, but there are often better ways to accomplish these functions with capacitors and active circuits.

It is the stray inductance that causes problems in high-speed designs. There is enough inductance in a short length of component lead or PCB trace to cause really significant potential problems in a circuit. We will see later that (just like capacitive coupling) inductive coupling between traces must often be carefully controlled or minimized.

CHARGE AND DISCHARGE CURRENTS

There are a few waveforms and relationships that people in our industry should be able to recognize and sketch from memory. Figures 3-10 through 3-12 represent some of these. Figure 3-10 illustrates a simple resistor-capacitor (RC) circuit. When the switch is one position, current flows toward the capacitor. When the switch is in the other position, the capacitor discharges.

When we first connect the battery, current flows toward the capacitor. In fact, from Ohm's Law, if R = 1.0 the flow will be 1.0V/1.0 Ω = 1.0 amp. Remembering that for capacitors, current leads voltage, and since the capacitor initially has no charge on it, even though the current flow is 1.0 amp, there will be 0.0 volts across the capacitor. Therefore, 1.0 volt must appear across the resistor (1.0 amp through 1.0 Ω).

As current flows onto the plate of the capacitor, the voltage across the capacitor must begin to rise. Since the battery voltage does not change, the voltage across the resistor must necessarily decrease (the sum of the voltages across the capacitor and across the resistor must be constant and, in this case, equal to the voltage of the battery, 1.0 volt). Since the value of the resistor does not change, then the current through it must begin to decrease. After some period of time, the current will decline to zero and the voltage across the capacitor will rise to equal that at the battery.

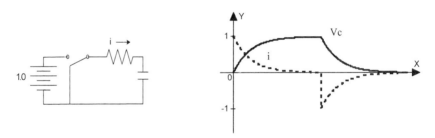

Figure 3-10 Typical charging and discharge voltage and currents for an RC circuit.

The graph in Figure 3-10 illustrates the relationship. As voltage across the capacitor increases, current decreases. All RC circuits follow a curve such as this, and differ only in terms of the height of the curves or how "stretched out" they are along the horizontal axis. But the shape of the curve is always that of an exponential, and we will define the equation for it in Chapter 4 in the section on time constants.

When we flip the switch to remove the battery from the circuit, the capacitor then discharges through the resistor. Note that the fully charged capacitor (1.0 volt) now causes a current of 1.0 amp (initially) to flow through the resistor in the opposite direction. Thus, the voltage across the resistor is now of the opposite sign. As before, the curves follow a very well-defined pattern. In fact, in this type of situation (a fixed DC source, a simple discharge path around the battery, and time to reach equilibrium) the discharge curves must be identical mirror images of the charging curves.

Figure 3-11 shows the corresponding RL (resistor-inductor) circuit. The analysis is identical. When the battery is first connected, the voltage across the inductor jumps up to 1.0 volt and there is no current through the inductor. Then as current begins to flow through the inductor, the voltage across the inductor begins to decline. At equilibrium, there is no voltage across the inductor, and the current is defined by the battery and the resistor:

$$i = V/R = 1.0/1.0 = 1.0 \text{ amp}$$

After the switch is thrown, disconnecting the battery from the circuit, the inductor tries to maintain the same current flow. Since the current doesn't change instantaneously, the voltage across the resistor doesn't change. So the voltage remains at 1.0 volt across the resistor. The left side of the resistor is now the same node as the bottom of the inductor (since the switch has switched), so the voltage across the inductor is now the opposite of what it was when the switch was first connected to the battery.

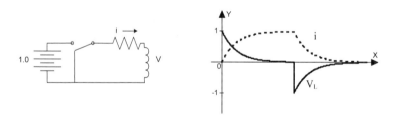

Figure 3-11 Typical charging and discharge voltage and currents for an RL circuit.

Again, these curves follow a well-defined exponential relationship, and the discharge curves are mirror images of the charging curves.

When a sine wave current source is applied to a capacitor or an inductor, the voltage across the device also follows a sine wave. For pure capacitors and inductors (i.e., no resistance component at all) the voltage waveform will be shifted (exactly) 90 degrees from the driving current waveform. (Remember, there are 360 degrees in one cycle.) This result can be proven mathematically, but the proof requires mathematical analyses well beyond the scope of this book. So you'll just have to take this fact on faith.

For capacitors, current will lead voltage by 90 degrees (Figure 3-12a) and for inductors it lags voltage by 90 degrees (Figure 3-12b). Conversely, for capacitors, voltage lags current by 90 degrees and for inductors voltage leads current by 90 degrees. If we connect a capacitor and an inductor in series, current must necessarily be the same through both of them (current is the same everywhere along a loop). Therefore, *the voltage across the capacitor would lag the voltage across the inductor by 180 degrees* (or exactly one-half cycle). Later on we will look at the total voltage across the inductor–capacitor (LC) series pair and note that since their voltages are 180 degrees out of phase, one subtracts from the other. A particularly interesting case is when the voltages across each component are equal, but 180 degrees opposite in phase. Even though there can be voltage across each component, the voltage across the pair will be equal and opposite at all times. Therefore, the combined voltage across the pair will be zero! This surprising result can have very important consequences for certain types of circuits.

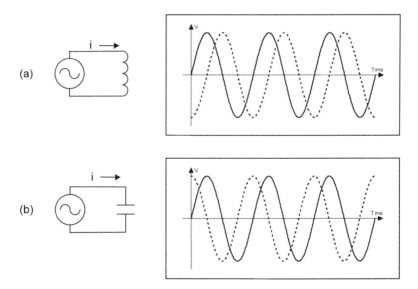

Figure 3-12 Voltage (solid line) leads current through an inductor (a) and voltage lags current through a capacitor (b).

Resonance

Before we leave our hydraulic analogies, let's look at one more special case. Suppose we have the storage tank (capacitor) and paddle wheel (inductor) in parallel as shown in Figure 3-13. Suppose we turn on the pump long enough to have the paddle wheel turning freely (no resistance to the flow). Then, let's turn off the pump.

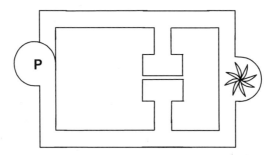

Figure 3-13 Hydraulic analogs for inductance and capacitance connected in parallel.

If the conditions are just right, inertia from the paddle wheel will cause water to continue flowing up into the lower storage tank. This will continue until the pressure in the tank gets strong enough to slow, and then stop, the paddle wheel. But when the paddle wheel finally stops, there will be considerable pressure in the tank. This will cause the paddle wheel to start turning in the opposite direction, moving water from the lower storage tank to the upper storage tank. This will continue until the pressure in the upper storage tank becomes strong enough to slow, and then stop, the paddle wheel. By that time, the pressure in the upper tank will be strong enough to start the paddle wheel turning in the opposite direction again, moving water from the upper storage tank to the lower one.

In the absence of friction (resistance), and if the tanks and paddle wheel are sized correctly, this can go on indefinitely! It requires no outside force or power to keep it going once it starts. (Of course, in the real world there is always friction, or resistance, but electronic circuits can be designed where these oscillations can be very large and strong.) If one component is very much larger or very much smaller than the other, then this oscillation will die out very quickly. But if the components happen to be sized just right, oscillations can continue for a very long time. This very special set of conditions is called *resonance*, and as we will see, it is a very important concept.

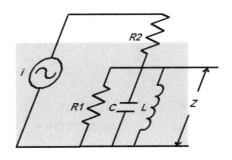

4 VOLTAGE AND CURRENT CHANGES AND TIME CONSTANTS

VOLTAGE AND CURRENT CHANGES THROUGH RESISTORS

*I*f current through a resistor changes, voltage across the resistor changes immediately. There are no time lags for voltage or for current through resistors. Therefore, the *impedance* that a resistor offers to the flow of current is solely a function of the size of the resistor. Frequency does not play a role. This is one of the biggest distinctions between resistors and the other two types of components.

If we plot the relationship for either voltage or current for a resistor over time, it is a straight line. For example, Figure 4-1 plots current through time for a fixed voltage applied across three different resistors. I do that here simply as a comparison against plots that will follow for capacitors and inductors.

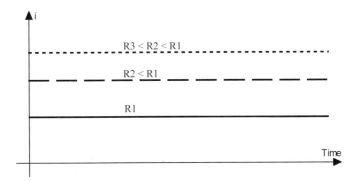

Figure 4-1 Current through a resistor through time.

Voltage and Current Changes Through Capacitors

Earlier, we defined capacitance by saying that a one-farad capacitor with one coulomb of charge stored on its plates will have one volt across it. Therefore, the voltage across a one-farad capacitor would rise one volt for each additional coulomb of charge flowing onto its plates. It follows that the voltage across a 2.0-farad capacitor would rise only half as much (.5 volts) for the same flow of charge. So we are talking about a change of charge producing a change of voltage for any given value of capacitance. Using the symbol, Δ, for change, we can state that

$$\Delta V = \Delta Coulombs/C$$

Since current is the flow of electrons (coulombs) during some measure of time, we can state that:

$$i = \Delta Coulombs/\Delta t(ime), \text{ or}$$
$$\Delta Coulombs = i \times \Delta t$$

Therefore, the voltage across a capacitor is related to current and to capacitance according to the fundamental relationship shown in Equation 4-1:

$$\Delta V = i \times \Delta t/C, \text{ or} \qquad [4\text{-}1]$$
$$i = C \times \Delta V/\Delta t, \text{ or}$$
$$C = i \times \Delta t/\Delta V$$

Expressed in more conventional (calculus) terms, substituting d for Δ, the relationships are shown in Equation 4-2:

$$dV = i \times dt/C, \text{ or} \qquad [4\text{-}2]$$
$$i = C \times dV/dt, \text{ or}$$
$$C = i \times dt/dV$$

Thus, the voltage buildup across a capacitor depends (a) on the value of capacitance (larger capacitors will build up voltage more slowly) and (b) how large a current flows for (c) how long a time. This actually is fairly intuitive. Larger capacitors build up voltage more slowly. Larger currents charge capacitors faster. Even small currents will charge up a capacitor over time. Figure 4-2 illustrates how the voltage will build up on a capacitor as a function of time and current level. If we are pulling current out of a capacitor, the same relationship exists. The change (decrease) of voltage will be a function of how much current we drain off for how much time for any given value of capacitance.

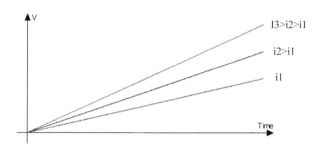

Figure 4-2 Curves showing voltage across a capacitor as a function of time for three different current levels.

Note that for any given level of current, i, even very large levels of current, the change in voltage is small for *small* periods of time. This illustrates the fact that voltage lags current, or, alternatively, current leads voltage with capacitors.

As an example, suppose we could charge a capacitor with a constant 10 mA of current for 1.0 microsecond. Table 4-1 illustrates the results for three values of capacitance. There will be almost no voltage built up across a relatively large (1.0 uF) capacitor, but almost 10 volts will build up across a .001 uF capacitor in the same time with the same current. This is one indication of why we want to use relatively larger capacitors as filter capacitors in power supplies.

Suppose we charge a capacitor with 10 mA for 1 us $\Delta V = i \times \Delta t/C$ $=10(10^{-3})(10^{-6})/C$			
if	$C = 1 \quad \mu F = 10^{-6}$	$\Delta V = .01$	volt
	$C = .01 \quad \mu F = 10^{-8}$	$\Delta V = 1.0$	
	$C = .001 \quad \mu F = 10^{-9}$	$\Delta V = 10.0$	

Table 4-1 Voltage buildup on a capacitor depends on current, time, and capacitance value.

In a typical bypass capacitor application, we are both charging and discharging the capacitor, depending on the switching cycle of the device being bypassed. The desire is to hold the voltage as constant as possible. It would follow, therefore, that larger capacitors would be better for bypass applications. Yet we often use smaller capacitor values in bypass applications. We will learn later that, whereas larger values appear to be preferable, lead inductance associated with bypass capacitors has some impact when selecting the appropriate values for bypass capacitors. It turns out that we sometimes compromise between capacitor value and the magnitude of lead (and connection) inductance.

VOLTAGE AND CURRENT CHANGES THROUGH INDUCTORS

It is the very nature of inductance that a constant current flow through an ideal inductor has no effect (except that it does generate a constant magnetic field around the inductor). But if we *change* the flow of current, then the magnetic field around the inductor changes and, as a consequence, a voltage is generated across the inductor. So, voltage across an inductor is a function of the *change* (Δ) in current flow. For a given change in current, the voltage generated is higher for larger values of inductance. And, the voltage level is higher if the current flow changes quickly than it is if the current flow changes more slowly.

This relationship is captured in Equation 4-3:

$$V = L \times \Delta i/\Delta t, \text{ or} \quad\quad [4\text{-}3]$$
$$L = V \times \Delta t/\Delta i, \text{ or}$$
$$\Delta i = V \times \Delta t/L$$

Expressed in more conventional (calculus) terms, substituting d for Δ, these relationships become those shown in Equation 4-4:

$$V = L \times di/dt, \text{ or} \quad\quad [4\text{-}4]$$
$$L = V \times dt/di, \text{ or}$$
$$di = V \times dt/L$$

If we apply a sudden voltage across an inductor, no current will flow during the first instant of time. (Refer to Appendix B for a thorough discussion of why inductors do what they do, i.e., induct.) As time goes by, more current will flow (see Figure 4-3). A higher voltage will cause the magnitude of the current to increase more quickly, and a larger inductor will cause the magnitude of the current to increase more slowly. But we have to change voltage *first* before current changes. As a result, we say that voltage leads current for inductors, or, alternatively, current lags voltage.

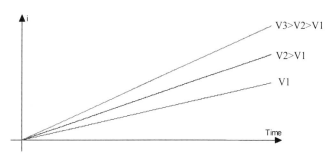

Figure 4-3 Curves showing current through an inductor through time for three different voltage levels.

One thing that is sometimes difficult for students to grasp initially is that, for a given voltage, current will increase linearly (as a function of time) forever! Obviously, most voltage sources can't supply infinite current, so the relationship shown in Figure 4-3 works for current levels that a source *can* supply and still hold voltage constant. In real-world applications, there comes a point where the source can no longer (both) supply increasing current and hold the voltage constant at the same time. Typically, the voltage begins to fall. This causes a reduction in the increase of current flow, and pretty soon a level of stability is reached at some level of current with no voltage across the inductor.

Note again that the instantaneous voltage developed across an inductor is given by Equation 4-4:

$$V = L\,di/dt$$

Table 4-2 shows why the di/dt term can be so important. If we change current by one amp in one second, the rate is equal to 1 amp/second. But if we change the current by only one mA over just one ns, the instantaneous rate of change is

1,000,000 (10^6) amp/seconds. It is this effect that is at the heart of so many high-speed design noise problems.

If di =	And dt =	Then di/dt =	Units
1 amp	1 second	1	amp/second
10^{-3} amp	10^{-3} second	1	amp/second
10^{-6} amp	10^{-3} second	10^{-3}	amp/second
10^{-6} amp	10^{-6} second	1	amp/second
10^{-3} amp	10^{-9} second	10^6	amp/second

Table 4-2 The di/dt term can become significant when the dt term is very small.

For example, suppose we have a lead inductance of 10 nH (10^{-8}) and we switch 10 mA (10^{-2}) of current through it. If we switch the current over 1.0 usec (10^{-6}), the instantaneous voltage generated across the lead inductance will be:

$$V = L di/dt = (10^{-8})(10^{-2})/(10^{-6}) = 10^{-4} = 0.1 \text{ mV}$$

But, under identical conditions, if we simply switch the current faster, at say 1.0 ns (10^{-9}) then the induced voltage will be:

$$V = L di/dt = (10^{-8})(10^{-2})/(10^{-9}) = 10^{-1} = 100 \text{ mV}$$

This latter voltage, 1,000 times higher than in the former case, might cause a noise problem, whereas the former case probably did not. Yet the *only* difference in these two illustrations is the change in time (rise time).

Some people have been unexpectedly "burned" when their integrated circuit (IC) suppliers "improved" their devices so that they switched more rapidly. Switching current rise times (di/dt) that were not a problem before suddenly become a problem, causing noise problems to occur where they didn't before. If you are not aware that your supplier has "improved" your circuits, or if you change suppliers of a particular IC and the new source switches faster, you might have a new problem and not have a clue how it happened.

Chapter 4

SOME INTERESTING INDUCTIVE CIRCUIT DYNAMICS

Figure 4-4 illustrates a circuit design characteristic that has caught more than a few engineers by surprise, much to their expense. The illustration shows a 10-volt source energizing an electromagnetic relay through a 100-Ω resistor. The transistor switch, Q, activates the relay. This is a very common type of circuit that is found, for example, in a variety of relay switching, scoreboard, and display applications. As shown, when the relay is activated, 100 mA (10V/100 Ω) will flow through the inductive coil of the relay.

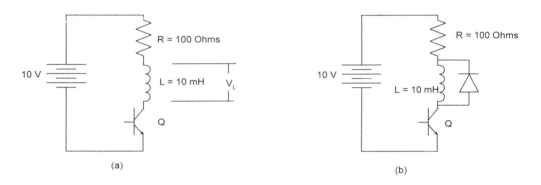

Figure 4-4 Reverse voltage protection diode (b) can protect a switching transistor (Q) from damaging voltage transients.

Now, let's turn off the relay in Figure 4-4a by turning off the transistor switch. And let's say the switch turns off in one microsecond. The inductive action (recall the inertia analogy) of the inductive relay coil will try to keep the current flowing. What happens is the magnetic field around the coil starts to collapse. This will generate a voltage, which will appear across the transistor. The instantaneous magnitude of the voltage will be, from Equation 4-4:

$$V = L\,di/dt$$

In this illustration, L is 10 mH (10^{-2}), di (the change in current) is 100 mA (10^{-1}) and dt (the change in time) is one microsecond (10^{-6}). So, the instantaneous voltage generated at the coil is

$$V = (10^{-2})(10^{-1})/(10^{-6}) = 1{,}000 \text{ V}.$$

The polarity of this voltage is positive at the junction of the collector of the transistor and the coil and negative at the other end of the coil. That is, the circuit is attempting to *maintain* the flow of current in the same direction as it was flowing before the transistor turned off.

This voltage exists only for a very short time (on the order of a microsecond), and if you saw it on a scope at all, it would appear as a very short voltage spike. But this voltage appears across the transistor. Transistors tend to be more tolerant of current spikes than voltage spikes (the current here is only 100 mA). After all, voltage spikes as small as tens of volts can puncture through the doped regions of a transistor, ruining it. More perversely, voltage spikes sometimes only weaken a transistor, so that the device may not fail until the voltage spike has occurred a large number of times (i.e., after the product has been in the field for a while).

This design flaw is, unfortunately, more common than you would think. If the circuit designer has forgotten (or maybe never learned) about inductive effects, it may go undiagnosed for a long time.

The solution is quite simple. Place a diode across the coil as shown in Figure 4-4b. This is called a *reverse voltage* diode in that it provides reverse voltage protection for the transistor. The diode provides a path for the current that the inductor tries to keep flowing, so that a voltage never has a chance to build up. The maximum current is only the 100 mA flowing through the coil, so the diode can be a fairly routine one (no special requirements).

This fairly routine circuit illustrates how destructive inductive effects (L x di/dt) can be in a design, and why an understanding of inductors and their effects is important for a circuit designer.

There is another fairly common circuit that depends on this same type of inductive effect for operation. It is the old automotive ignition system shown in Figure 4-5. When the points are closed, current flows from the 12-volt car battery through the coil. At the right moment, the points open and the current stops flowing, causing the magnetic field around the coil to collapse. The automotive ignition coil is also a transformer, so the collapsing field around the entire coil causes a very high voltage to appear across the spark plug. A high voltage also appears across the points, which can cause them to burn out. That is why older ignition systems had "condensers" (really capacitors) around them to prevent the points from arcing and burning. Modern cars still have coils and they generate spark across plugs in much the same way. But the points have been replaced by improved electronic switching, which is (a) faster (shorter dt), so higher spark voltages can be generated, (b) more reliable, and (c) requires less maintenance (no mechanical points to replace).

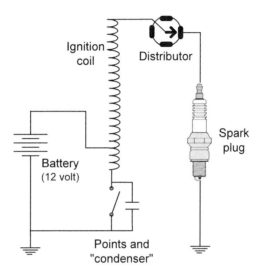

Figure 4-5 Early automotive ignition systems used points in the spark generation system.

TIME CONSTANTS

Let's look at Figure 4-6. Recall from earlier that when the switch connects the resistor to the battery, current will immediately begin to flow onto the plates of the capacitor. The initial magnitude of the current (assuming the capacitor is initially uncharged) is V/R. Voltage across the capacitor will begin to increase according to Equation 4-1:

$$\Delta V = i \times \Delta t / C$$

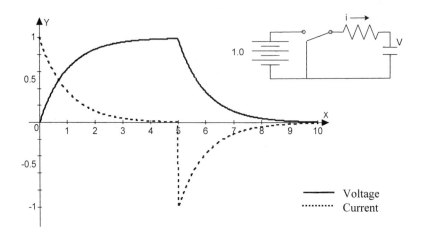

Figure 4-6 Charging and discharge curves associated with RC circuits.

As the voltage across the capacitor increases, the voltage across the resistor must decrease, since the battery voltage remains unchanged. If the voltage across the resistor decreases, the current through the resistor must also decrease. Thus, the current starts out at the value V/R but decreases steadily through time until it reaches zero (when the capacitor is fully charged). Similarly, the voltage across the resistor starts out at V, but declines steadily through time until it reaches zero (when the current stops flowing because the capacitor is fully charged). Finally, the voltage across the capacitor starts out at zero, but grows steadily until it reaches the value V.

The curves these voltages and currents follow are shown in Figure 4-6. For basic RC circuits such as the one shown, they follow well-defined, predictable shapes. When the capacitor is charging, the equations for the curves are given by Equation 4-5 and 4-6.

Voltage across the capacitor:

$$VC(t) = V(1 - e^{-t/RC}) \quad\quad\quad [4\text{-}5]$$

Current through the capacitor:

$$i(t) = (V/R)e^{-t/RC} \quad\quad\quad [4\text{-}6]$$

The value e is a constant (e = 2.71828...) and it is the base of the natural logarithm (see Appendix C). These types of expressions are called *exponential equations* and they are very important in electrical engineering. If t equals zero then the exponent of e also equals zero, regardless of the values of R and C. Any value raised to the zero power has a value of one (e.g., $e^0 = 1$). So the initial conditions of these equations are

$$VC(t = 0) = V(1 - e^{-0}) = 0$$

$$i(t = 0) = (V/R)e^{-0} = V/R$$

as we would expect. If any value is raised to a very large negative value (e.g. e^{-n}, where n ≅ infinity) its value becomes zero. So, for very large values of t (after the circuit stabilizes), the equations become:

$$VC(t) = V(1 - 0) = V$$

$$i(t) = (V/R)0 = 0$$

We could, if we wanted to, generalize these equations in units of RC. That is, we can look at the equations at the points where t = 1RC, t = 2RC, ..., t = nRC. To do so, lets think of a unit of time called tc where tc = 1 if t = 1RC, tc = 2 if t = 2RC, and so on. If we do that, the preceding equations become generalized as Equations 4-7 and 4-8.

$$VC(t) = V(1 - e^{-tc}) \qquad [4\text{-}7]$$

$$iI(t) = (V/R)e^{-tc} \qquad [4\text{-}8]$$

Interestingly enough, these equations now can apply to *any* circuit with a resistor and a capacitor of *any* arbitrary values, as long as we recognize that tc = t/RC. The expressions e^{-tc} and $(1 - e^{-tc})$ are graphed in Figure 4-7. They illustrate some interesting characteristics. For example, at a time equal to 1tc, the curves are 63% of the way toward their final value. At a time equal to 3tc they are 95% of the way to their final value. After a time of 5tc they are virtually at their final value (for all practical purposes).

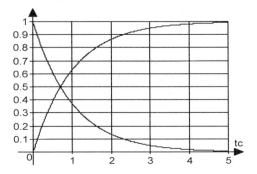

tc	e^{-tc}	$1 - e^{-tc}$
1	.37	.63
2	.14	.87
3	.05	.95
4	.02	.98
5	.01	.99
6	.00	1.00

Figure 4-7 Exponential curves expressed in "time constants."

The value RC (the value of R times the value of C) is called an *RC time constant* and is an important concept. The response of *any* circuit with a resistance and a capacitance will follow the shape of an exponential curve and will reach 63% of its final value in a time equal to R x C. It will reach 95% of its final value within three RC time constants. For all practical purposes it has stabilized within 5 time constants.

This has both good and bad implications. For example, a stable resistor combined with a stable capacitor can be (and is routinely) used as a very accurate timing circuit. If we combine a 1.0-K resistor with a .001-uF capacitor in a timing circuit, the voltage across the capacitor will reach .95 V in 3RC or $3(10^3)(10^{-9}) = 3$ μsec. If the resistor and capacitor are good quality and stable, this will *always* be true for the life of the product. RC timing circuits are routinely used in products ranging from home appliances to cameras to audio sound systems to industrial systems to spacecraft.

On the other hand, the relationships also show that a circuit requires some time to respond. AC circuits usually cannot respond instantaneously. If we have an IC that is switching an output voltage through a 50-Ω output resistance into another IC with 20-picofarads (pF) input capacitance, it will take $3RC = 3(50)(20)(10^{-12}) =$ 3 ns to reach 95% of its value at the input of the receiving IC. That delay may not be fast enough for certain high-speed circuits.

An inductive circuit works in a very similar way. Figure 4-8 shows a resistor and an inductor in series. When the switch first closes, no current flows because of the inductive inertia of the inductor. If there is no current, there is no voltage across the resistor, so all the voltage appears across the inductor. As soon as some current flows, voltage is developed across the resistor, so the voltage across the inductor begins to decrease. The shapes of the curves for this circuit are identical to the shapes of the curves for the RC circuit, except:

1. The voltage across the inductor starts out high and declines, whereas the current through the inductor starts out low and increases,

2. The voltage across the capacitor starts out low and increases, whereas the current through the capacitor starts out high and decreases.

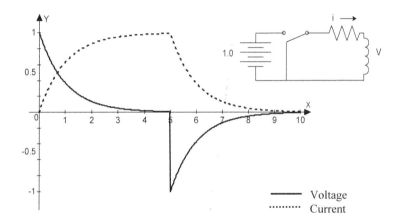

Figure 4-8 Charging and discharge curves for RL circuits.

This is another example in which capacitors and inductors are nearly equal but opposite in their effects.

The formulas for the initial charging curves for the inductor are similar to those for an RC circuit and are given by Equations 4-9 and 4-10:

$$i(t) = (V/R)(1 - e^{-tR/L}) \qquad [4\text{-}9]$$
$$V_L(t) = (V)e^{-tR/L} \qquad [4\text{-}10]$$

Just as we developed an explanation for an RC time constant earlier, we can think about an L/R time constant here. The circuit will reach 63% of stability in a length of time equal to L/R and 95% of stability in a length of time equal to 3L/R.

All of the strengths and weaknesses associated with an RC time constant are also true for an L/R time constant. We usually do not use inductors in timing circuits, however, because they tend to be large, bulky, and (often) not as stable as capacitors. In general, RC circuits perform better as timing circuits. (Note: We will leave the discussion about LC timing and frequency control for later.)

A Note on Charge and Discharge Equations

The discussion here has focused on charging circuits, when we switch the circuit to the battery. The description of what happens when we switch the circuit back to discharge the capacitor or inductor follows the same logic. The charging and discharge equations are summarized in Table 4-3. Note that the capacitor discharge current flows in a direction opposite to that of the charging current; hence the negative sign. When the magnetic field around the inductor begins to collapse, it tries to maintain the current flow around the loop. The voltage across the resistor (initially) still is V. So the voltage across the inductor reverses (hence the negative sign) to try to maintain the current flow. That also helps explain why the reverse voltage protection diode is oriented the way it is.

Table 4-3 Summary of RC and RL charging and discharge equations.

Circuit	Parameter	Charging	Discharging
RC	Vc	$Vc(t) = V(1 - e^{-t/RC})$	$Vc(t) = Ve^{-t/RC}$
	i	$i(t) = \left(\dfrac{V}{R}\right) e^{-t/RC}$	$i(t) = -\left(\dfrac{V}{R}\right) e^{-t/RC}$
RL	VL	$VL(t) = Ve^{-tR/L}$	$VL(t) = -Ve^{-tR/L}$
	i	$i(t) = \dfrac{V}{R}(1 - e^{-tR/L})$	$i(t) = \dfrac{V}{R}(e^{-tR/L})$

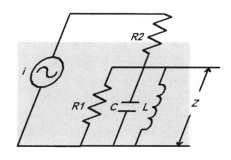

5 RESISTANCE

People who study electronic circuit analysis use a lot of calculus. Other than the basic di/dt term I've already introduced, I promise to stay as far away from calculus as I can in this book, but we can't stay away from equations. The whole point of selecting component values in circuits is to obtain a desired result. And the primary way you select component values is to perform calculations.

This book is not intended to show you how to make all the calculations. If you want to know that, you should go get your degree. But this book will cover some of the more fundamental calculations related to electronic circuits. And it will suggest what some of the more difficult calculations are, without actually trying to teach you how to do them. The objective is to give you enough familiarity with the topic that you can do some of the more basic calculations yourself, and can understand some of the more complex ones when someone does them for you.

KIRCHHOFF'S LAWS

Gustav Robert Kirchhoff is credited with (among many other things) two laws related to how currents flow in circuits. These two laws are almost as important to the field of electronics as Ohm's Law is. Consider Figure 5-1a and Figure 5-1b.

Figure 5-1a illustrates the case of two components (they don't have to be resistors) connected in parallel and driven by a voltage source (which could be DC or AC). Current will flow from the source and divide between the two (or more) components. The point where the voltage source and the two components come together is called a *node*. (This is equivalent to a net in PCB design terms.)

Kirchhoff's 1st Law states that the current flowing into a node must equal the current flowing out of the node. This really makes intuitive sense; it is a statement of conservation of charge. Since current is the flow of electrons, if the same number of electrons are not flowing into and out of a node, where do the excess ones come from or go to?

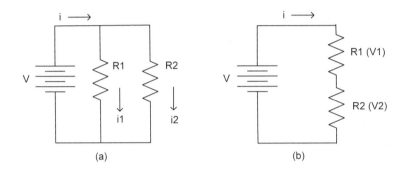

Figure 5-1 Two resistors connected in parallel (a) and in series (b).

Figure 5-1b illustrates two components connected in series and driven by a voltage source (AC or DC). It should be intuitively clear that the voltage across the series components must equal the voltage source. If not, how is the extra voltage generated and where does it go? Kirchhoff's 2nd Law states this in a form that students sometimes find a little awkward at first. The law states that the sum of the voltages around a loop must be zero. This is a conservation of energy law. Consider how to explain the case where they might not sum to zero.

The way to interpret Kirchhoff's 2nd Law is to start at the top and move clockwise around the loop. The first component has a voltage drop across it of V1. The second component has a voltage drop of V2. So far, the sum is V1 + V2. Continuing around the loop we come to the voltage source. The voltage at the voltage source is *minus* V. It is minus because we are approaching it (measuring it) from the bottom to the top as we continue to travel around the loop. So now we have Equation 5-1:

$$V1 + V2 + (-V) = 0, \text{ or}$$
$$V1 + V2 = V \qquad [5\text{-}1]$$

The trick to applying Kirchhoff's 2nd Law is to remember that the direction you are traveling around the loop determines the sign of the voltages you are considering.

SERIES RESISTORS

There are numerous times in circuits where we want to replace two or more resistors with an equivalent single resistor, or, occasionally, replace a single resistor with two of them. Thus, we think in terms of circuit equivalents. In this example, the question is, "What single resistor is the equivalent of two resistors connected in series?"

Most of us already know the answer. But the point is not in knowing the answer; the point is knowing how to do the analysis to get the answer, because later on we will ask the same question about equivalent capacitors and equivalent inductors. Those analyses will be very similar to the analysis we go through here. Although many people can talk about equivalent resistive circuits, they really don't understand similar equivalent circuits when we get to the other types of components.

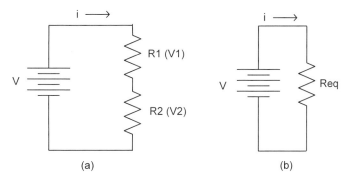

Figure 5-2 Equivalent circuit (b) of two resistors connected in series (a).

Figure 5-2a illustrates the case in which we are going to determine the single resistance, Req, that is the equivalent of the two resistors, R1 and R2, in series. First, recall from Kirchhoff's 2nd law that the sum of the voltages around a loop must equal zero. If V1 is the voltage across R1 and V2 is the voltage across R2, then if the sum of the voltages around the loop (from Equation 5-1) must equal zero:

$$V1 + V2 - V = 0, \text{ or}$$

$$V = V1 + V2$$

Also, if Veq is the voltage across the equivalent resistor, then the same analysis leads to Equation 5-2:

$$Veq - V = 0, \text{ or}$$

$$V = Veq \qquad [5\text{-}2]$$

Remember, as we go around the loop clockwise, we see the negative side of the battery first. So, going around the loop, the voltage across the battery is $-V$.

Remember, also, that the current around a loop is constant at all points in the loop. Let the current here be designated as i. The voltage across each resistor is (from Ohm's Law) $V = iR$. So we can write Equations 5-1 and 5-2, respectively, as Equation 5-3:

$$V = iR1 + iR2, \text{ and}$$

$$V = iReq \qquad [5\text{-}3]$$

This leads to:

$$iReq = iR1 + iR2$$

and dividing through by i, to Equation 5-4:

$$Req = R1 + R2 \qquad [5\text{-}4]$$

This might seem to be a lot of trouble to get to something that is somewhat obvious. But before we make that judgment, let's do a similar thing to parallel resistors. Then, wait until you see how this applies later to the combination of capacitors.

Chapter 5

PARALLEL RESISTORS

Figure 5-3 shows two resistors connected in parallel. Now the question is, "What value of resistor, Req, is equivalent to the two resistors, R1 and R2, connected in parallel?" Already the situation is going to get a little more interesting.

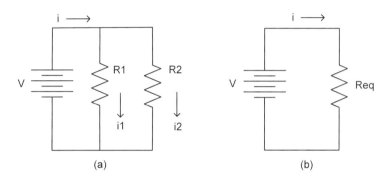

Figure 5-3 Equivalent circuit (b) of two resistors connected in parallel (a).

Kirchhoff's 1st Law states that the current going into a node has to equal the current going out of a node. This example is actually pretty simple, with just a single node at the top and another at the bottom. The current going into the node at the top is i. The current through R1 is i1 and the current through R2 is i2. By Kirchhoff's 1st Law we can write Equation 5-5:

$$i = i1 + i2 \qquad [5\text{-}5]$$

In this example, there are three loops: (a) the loop from the battery through R1, (b) the loop from the battery through R2, and (c) the loop between the two resistors, R1 and R2. By Kirchhoff's 2nd Law, the voltages around a loop must sum to zero. So we can write the following three equations to illustrate this:

$$i1R1 - V = 0 \qquad [5\text{-}6]$$
$$i2R2 - V = 0 \qquad [5\text{-}7]$$
$$i2R2 - i1R1 = 0$$

These are not really three different equations. Any one of these is a necessary result of the other two. We actually have only two independent equations here, and the third one is redundant.

We can write Equations 5-6, 5-7, and also 5-3 as follows:

$$i1 = V / R1$$
$$i2 = V / R2$$
$$i = V / Req$$

Then, substituting into Equation 5-5

$$\frac{V}{Req} = \frac{V}{R1} + \frac{V}{R2}$$

And, dividing through by V we get Equation 5-8:

$$\frac{1}{Req} = \frac{1}{R1} + \frac{1}{R2} \qquad [5\text{-}8]$$

which can alternatively be expressed as Equation 5-9:

$$Req = \frac{1}{\frac{1}{R1} + \frac{1}{R2}} = \frac{R1R2}{R1 + R2} \qquad [5\text{-}9]$$

The form of Equation 5-8 is important. By extension, it can be shown that if there are an arbitrary number of (n) resistors in parallel, the equivalent resistance is found from Equation 5-10:

$$\frac{1}{Req} = \frac{1}{R1} + \frac{1}{R2} + \text{---} + \frac{1}{Rn} \qquad [5\text{-}10]$$

which can alternatively be expressed as Equation 5-11:

$$Req = \frac{1}{\frac{1}{R1} + \frac{1}{R2} + ---+ \frac{1}{Rn}} \qquad [5\text{-}11]$$

These are relatively easy forms to use if it is ever necessary to calculate the equivalent resistance of several resistors in parallel.

It is interesting to note what happens in a few special cases. Suppose R1 is zero. The equivalent series resistance is still R2, but the equivalent *parallel* resistance is zero. If R1 is infinite, the equivalent *series* resistance is infinite, but the equivalent parallel resistance is R2. Thus, it follows that:

1. In the case of resistors in series, the resistance is never less than the largest resistor.

2. In the case of parallel resistors, the resistance is never more than the smallest resistance.

Also, if there are n resistors in parallel, and they are all the same value (R), the equivalent resistance is

$$Req = R/n$$

This is very handy to remember when we have multiple loads, for example, at the end of a trace.

VOLTAGE DIVIDERS

One of the more common resistive circuit configurations is the voltage divider (Figure 5-4). Designers should be able to recognize this particular configuration and the principle behind it.

A voltage source, Vout (from a previous stage in the circuit) is connected to two resistors in series. Recognize that the current is:

$$i = Vout / (R1 + R2)$$

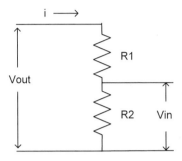

Figure 5-4 Classic voltage divider circuit.

and therefore we obtain Equation 5-12:

$$Vout = i(R1 + R2) \qquad [5\text{-}12]$$

The input (to the next stage) voltage, Vin, is given by Equation 5-13:

$$Vin = iR2 \qquad [5\text{-}13]$$

So, the ratio of Vin to Vout is given by Equation 5-14:

$$Vin/Vout = R2/(R1 + R2) \qquad [5\text{-}14]$$

Or, put more simply, the ratio of the input voltage (to the next stage) to the output voltage (from the previous stage) is the same as the ratio of the resistance below the input to the total resistance.

Figures 5-5a and 5-5b illustrate two common voltage divider configurations. In Figure 5-5a the voltage input to the operational amplifier follows equation 5-14. Figure 5-5b illustrates a typical simple volume control application. In this case, Vin is Vout x R1/R, where R1 is the resistance between the wiper of the control and the end connected to ground.

This type of configuration, where the result is proportional simply to one resistance divided by a total resistance, is so common it would be helpful if a designer could readily recognize it.

Chapter 5

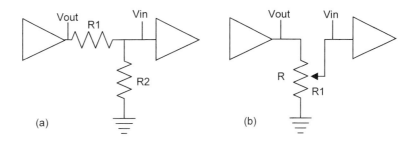

Figure 5-5 Two common voltage divider circuits.

AMPLIFIER FEEDBACK AND GAIN

A more subtle type of circuit will be included here simply for thoroughness. It is a common circuit, but sometimes the configuration is subtle enough that it might not be immediately recognized.

In Figure 5-6, V1out is a voltage signal being fed to the inverting input of an amplifier stage through resistor R1. V2out is the output of the next amplifier stage. Vv is the voltage at the input of the amplifier, so V2out is –Vv times the gain of the amplifier. Assuming the gain is very high, then it is true that

$$V2out \gg -Vv$$

It is also true that with a high-gain amplifier, whose other input is tied to ground (zero volts), Vv will be very close to zero; so close in fact that we can assume it does equal zero for simplicity. Since there is typically no current into the high-impedance amplifier, and since the current into a node must equal the current out of it, the current through R1 (i1) must be equal to the current through R2 (i2). If Vv is (virtually) zero, then the following relationships must hold true:

$$\begin{aligned} i1 &= i2 \\ V1out/R1 &= -V2out/R2 \\ V2out/V1out &= -R2/R1 \\ V2out &= -(R2/R1)V1out \end{aligned} \qquad [5\text{-}15]$$

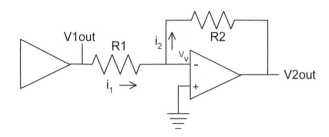

Figure 5-6 The amplifier gain of the second stage is R2/R1.

This is a very common relationship in analog design, where the gain of a stage is simply the (negative) ratio of the feedback resistor (R2) to the input resistor (R1). Sometimes, by looking at a schematic, it is not immediately obvious what a circuit designer has done. And there are many variations on this design concept. But when a circuit designer points out to you that the gain of a stage is simply the ratio of two resistances, you can now visualize why that is so.

POWER

We intuitively know that if we short-circuit the terminals of a typical 9-volt battery found in small electronic appliances, not much happens. But if you have ever seen an accidental short-circuit of the terminals of a 12-volt car battery, the result can be severe. (If you haven't seen it, don't try it!) The difference is that the car battery can supply a lot more power than the little 9-volt battery can.

Power is defined as voltage times current. A car battery that supplies 12 volts at 60 amps to a starter motor is supplying 720 watts of power (almost one horsepower: one hp = 746 watts). A 100-watt light bulb requires almost 1.0 amp at 110 volts. A 15-amp circuit breaker in your home will allow 1,650 watts (at 110 volts) before it trips. An alkaline D-cell (1.5-V) battery is typically rated for 100 hours at 150 mA., meaning it can supply 1.5 x .150 or almost .25 watt of power for that 100 hours.

Looking at the basic circuit in Figure 5-7, and remembering Ohm's Law (V = iR), we can express the formula for power these three ways:

$$\begin{aligned} \text{Power} &= \text{Vi} \\ &= V^2/R \\ &= i^2R \end{aligned} \qquad [5\text{-}16]$$

These equations reflect things we may intuitively recognize. For example, if we increase the voltage across a resistor, the resistor begins to get warmer. That reflects the increased power being dissipated within it (V^2/R). If we increase the amount of current through a PCB trace, the temperature of the trace increases (i^2R). If we put too much current through it, the trace can get so hot it melts. CMOS circuits run cooler than ECL circuits because, for the same voltage levels, they require much less current, and therefore much less power.

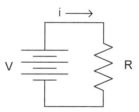

Figure 5-7 Basic circuit for illustrating power.

It is important to note that in all these cases, the power (and therefore the temperature) goes up as the *square* of the voltage or current. Thus, a 100-Ω half-watt resistor can handle 7 volts at .07 amps. But a 1-watt (twice as large) 100-Ω resistor can only handle 10 volts at .1 amps.

Most of the time, PCB designers don't have to worry much about power. The exception can be in power supplies, and this can really be a problem if we are designing large motherboards that supply lots of power to numerous daughter boards. If we are concerned about power, it is usually a concern about the i^2R (pronounced "eye-squared-R") power dissipation in traces. A discussion about i^2R power capabilities of traces is outside the scope of this book, but the reader can consult several articles on the subject, including the IPC trace/temperature tables.[1] There is also a PCB trace temperature calculator available for download from Ul-

---

[1]IPC-2221, *Generic Standard on Printed Board Design*, February 1998, Figure 6-4, p. 38.

traCAD's Web site at http://www.ultracad.com/calc. You are particularly encouraged to read the Help file that accompanies that calculator.

While the responsibility for power management usually belongs (first) to the circuit designer, board designers should understand the implications associated with i^2R losses and considerations on their boards and be alert to potential heating problems caused by high currents. The most common and effective ways to deal with heat range from (a) relocating parts (to avoid heat buildup at a localized area), (b) resizing parts (for example increasing the size of traces), and (c) improving cooling techniques (perhaps with fans, board orientation, heat sinks, or even metal "cores" within boards).

EQUIVALENT CIRCUITS

So far, all the circuit examples we have looked at have been fairly simple, mostly involving a single power source and one or two resistors. What if we have a more complicated example like that illustrated in Figure 5-8? How would we go about analyzing the output voltage of that circuit?

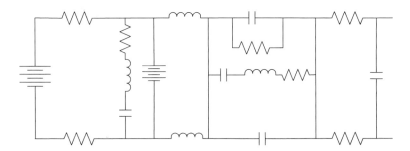

Figure 5-8 This complex circuit would be difficult to analyze.

Well, it can be done. The process would be a sequential one of taking a couple of components at a time and finding their equivalents, and then reducing the circuit almost component by component until it becomes manageable.

There are two very interesting theorems in electronics that conceptually make this a very simple task, at least in an empirical (experimental) sense. They go like this:

Any circuit, such as the one shown in Figure 5-8, can be reduced to an equivalent circuit containing a *single* ideal power source and a *single* resistor, as shown in either Figure 5-9a or Figure 5-9b.

Figure 5-9a illustrates what is known as a Thevenin equivalent circuit. It consists of an ideal voltage source, V, with a resistor, R, in series with it. Figure 5-8 can be reduced to Figure 5-9a using the following strategy: The voltage, V, is the open circuit voltage that would be measured at the output (right) side of the network if there were no other connections made there. The resistor, R, is determined by dividing this voltage by the short-circuit current that would be measured if you shorted the terminals at the right of the network. Thus, the measured open-circuit voltage of Figure 5-9a is the same as that of Figure 5-8. Also, the measured short-circuit current of Figure 5-9a is the same as the short-circuit current of Figure 5-8. Now, if you connected any resistor to the output of the equivalent circuit represented in Figure 5-9a, the current through it, and the voltage across it, would be the same as if you connected that resistor to the output side of the circuit in Figure 5-8. This is the power and the beauty of Thevenin's Theorem.

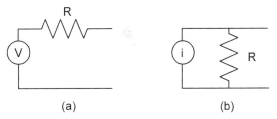

Figure 5-9 Thevenin (a) and Norton (b) equivalent circuits

Figure 5-9b illustrates a Norton equivalent circuit. It consists of an ideal current source, I (as opposed to the voltage source, V, in the Thevenin equivalent circuit), and a resistor, R, in parallel with the current source. In a manner similar to the preceding, the magnitude of the current source in Figure 5-9b is equal to the short-circuit current at the output in Figure 5-8. The resistor, R, is determined by dividing the open-circuit voltage at the output of Figure 5-8 by this short-circuit current. As with the Thevenin circuit, if you attach any resistor at the output of the equivalent circuit of Figure 5-9b, the current through it, and the voltage across it, will be the same as if you added it to the output of the actual circuit of Figure 5-8.

So what? Does this have any practical value?

I picked up a National Semiconductor data sheet at random. It happened to be for a DS26F31C/DS26F31M Quad High Speed Differential Line Driver. It, like Figure 5-8, has a lot of stuff inside it. But it has a typical "high" voltage spec of 3.2 volts and an output short-circuit spec of 60 mA. If we divide the voltage by the current (3.2/.06) we get 53 Ω. If we want to see how this device will work in a circuit (at least while its voltage is high) we can simply substitute Figure 5-9a or Figure 5-9b in its place in the circuit, with R = 53.3 and V = 3.2 (Figure 5-9a) or i = .06 (Figure 5-9b). We don't *need* to know all the circuitry inside it. All we need to know is the open circuit voltage and the short-circuit current. Virtually all data sheets will provide this information. The data sheet will also provide other general guidelines, for example, for the output capability of a device. But using the implied output resistance is a way to determine circuit performance for, perhaps, configurations that are a little bit different than standard.

As an aside, the value, in this case, for R, 53 Ω, is for each side of the differential output for this particular device. The differential output pair of this device is designed to be terminated with a 100-Ω resistor. It is not an accident that this happens to be twice the value for R we just determined. The significance of this relates to transmission line considerations, which we will cover in great detail in Chapters 10 and 11. There is a very good reason why the implied value for R we just determined is 53 Ω in this equivalent circuit, and not some other value.

The calculations we just made apply to the "high" output voltage. It might well be that the equivalent circuit will look different when the device is at its "low" state. That is one problem with analyzing the performance and effects of an active (semiconductor) circuit. Equivalent circuits are legitimate tools when ICs are in a "stable" state (at a "high" or a "low," or at a "one" or a "zero"). They usually don't apply during transient switching times between states. But we are usually concerned about different things then. At a stable state, we are concerned about voltage and current levels. During switching, we are usually concerned that the device can, in fact, *complete* the switching cycle in the appropriate time. We will see later that other device specifications (especially things like input capacitance and output rise times) are of interest during switching times.

Power Curve

Linear ICs can often be looked at from the standpoint of their equivalent circuits. For example, Figure 5-9a could easily apply to the output stage of a transmitter or to an audio amplifier. When we add a load to such circuits (such as an antenna to a transmitter or speakers to an audio amplifier) it is interesting to see what happens to output power as a function of that load.

Referring to Figure 5-10, let V be 5 volts and Rs (source resistance) be 50 Ω. RL is the load resistance. The short-circuit output current, if we let RL be zero, would be V/Rs = 5/50 = 100 mA. If we did short-circuit the outputs there would be 5 volts across, and 100 mA through the 50-Ω source resistor, Rs. Consequently, .5 watts would be dissipated in the resistor. It is important to note that this power is dissipated within the device. The power dissipated in the load outside the device would be i^2RL, but since RL is zero (short circuit), there is no power dissipated in the load.

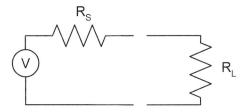

Figure 5-10 Voltage source powering a load, RL, through a source resistor, Rs.

If the load resistor, RL, were 50 Ω, then the output current would be:

i out = V/(Rs + RL) = 5/100 = 50 mA

There would be 2.5 volts across each resistor, and therefore .125 watt (2.5 volt x .05 amp) dissipated in each resistor. Now, the power dissipated within the device is only .125 watt, and the output power, the power in the output resistor, is also .125 watt. The total power has decreased, since the total resistance looking at the voltage source has increased, but the output power (the power dissipated in the load) has increased. Since power usually manifests itself as heat, the heat generated within the device is much less, as we would expect. Short-circuits are usually hard on devices.

Now let RL be 100 Ω. The total current is now:

$$i\ out = V/(Rs + RL) = 5/150 = 33\ mA$$

There is now .055 watts dissipated within the device (in Rs) and .11 watts in the load resistor, for a total power produced of .166 watts.

What has been illustrated here is a fundamental truth. Output power (the power dissipated in RL) is maximized when RL equals Rs (Figure 5-11). *Total* power decreases as RL increases, but the power in the *load* is a maximum when the load resistor is the same as the source resistor. This illustration has been with resistors for simplicity, but this relationship is true, in general, for any type of impedance. Output power is maximized when the load impedance equals the source impedance. When output power is maximized, an equal amount of power is dissipated within the device and in its load.

That is why we buy audio speakers to match the output impedance of our audio systems and why antennas are matched to transmitters. Even good CB radios have antenna-matching adjustments to help match the antenna load impedance to the output impedance of the transmitter. The reason is so that the available output power (the power dissipated in the load, i.e., in the speakers or the antenna) is maximized.

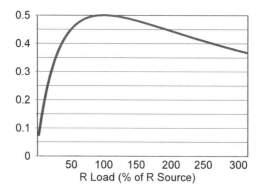

Figure 5-11 Output power is a maximum when the load impedance equals the source impedance.

Do we always want to maximize power? Of course not. Sometimes we want to maximize signal sensitivity. Then we would design input stages with higher levels of impedance so they don't "load down" their source (but not too high, because this might increase noise sensitivity). Sometimes we have to design low-impedance outputs for a device so that can drive other devices with lower impedance inputs. With transmission lines, we want to match impedances in order to reduce reflections (not to maximize power). But, sometimes we do want to maximize power transfer between devices, and in this case doing so requires that we match the load impedance to the source impedance.

POWER SOURCES

Up until now we have used pure, or ideal, energy sources in our examples. For example, Figure 5-12a shows a battery and a resistor. We introduced Ohm's Law showing that, for example,

$$i = V/RL$$

But what if RL is zero? That would imply an infinite current, which we know intuitively is not possible. The equivalent circuit for a battery (or more generally for any energy source) is more properly shown in Figure 5-12b, an ideal energy source with a source resistor, Rs, in series with it.

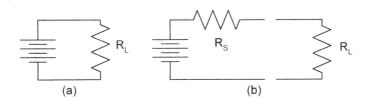

Figure 5-12 Battery voltage source shown as an ideal source (a), and as the more realistic case with source resistance (b).

A 12-volt car battery, for example, might be able to deliver a short-circuit current of perhaps 80 amps. That would imply a value for Rs of around 12/80 = .15 Ω. Now, when driving the starter motor, we want to deliver maximum current to the starter. If the cable between the battery and the starter motor had a resistance of just .15 Ω, we would cut the maximum available current at the starter by 50%. That is why battery cables are so big. Their resistance must be small compared to the source resistance, Rs, of the battery.

Looking at this another way, what would the battery cable resistance need to be if we only wanted to drop a maximum of 1.0 volt across the cable when drawing 60 amps to the starter motor? A resistor of 1.0V/60A = .017 Ω will develop a voltage across it of 1.0 volt at 60 amps, so the total battery cable resistance, including the connections to the battery and the started motor, must be just .017 Ω if we want to limit the voltage drop to 1.0 volt. (Keep those battery cable connections clean!)

A fresh alkaline D-cell battery can (perhaps surprisingly) deliver 10 amps of short-circuit current (at least for a few moments). So, the initial output resistance of a fresh battery is

$$Rs = 1.5 \text{ volts}/10 \text{ amps} = .15 \text{ Ω}$$

When the battery appears to be running down, the output resistance is going up. Therefore, it can deliver less and less current without pulling the output voltage down across the output resistance. A so-called "dead" battery has a large output resistance and drops virtually all the available voltage across that output resistance at almost any load.

CONDUCTANCE

Conductance is a term we don't very often hear used in general engineering discussions. It is defined as the inverse of resistance. If something offers high resistance to current flow, there will be very little conductance, and vice versa. Conductance is designated by the symbol G. Thus,

$$G = 1/R$$

The concept of conductance is primarily used to simplify certain kinds of mathematical analyses. For example, if we express Equation 5-11 in terms of conductances, it would be Equation 5-17:

$$R_{eq} = \cfrac{1}{\cfrac{1}{R1} + \cfrac{1}{R2} + \text{---} + \cfrac{1}{Rn}} = \cfrac{1}{G1 + G2 + \text{---} + Gn} \qquad [5\text{-}17]$$

We will not use conductance again in this book until we discuss lossy transmission lines in Chapter 17.

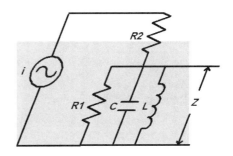

6 REACTANCE

*A*s I have indicated, there are only three types of passive components: resistors, capacitors, and inductors. Each of these *impedes* the flow of current, but each does so in a different way. We categorize them these three ways:

- *Resistance* (designated by the symbol R) is the impedance to current flow that is presented by a pure (ideal) resistor. Resistance is measured in ohms. Resistance is directly related to the size of the resistor, but it is not a function of frequency and there is no phase shift associated with this impedance to current flow.
- *Reactance* (designated by the symbol X) is the impedance to current flow that is presented by an ideal inductor or capacitor. Reactance is measured in ohms. Reactance is related to the size of the component but is *also* related to the frequency of the signal. There is a 90-degree (only) phase shift associated with reactance.[1]
- *Impedance* (designated by the symbol Z) is what we get when we *combine* resistance and reactance (see Chapter 7). Impedance is measured in ohms. It is a function of frequency and there is normally a phase shift associated with impedance. That phase shift can range between +90 degrees and –90 degrees.[2]

---

[1] Note that this statement says there is a 90-degree phase shift (plus or minus) associated with reactance. It is *exactly* +/–90 degrees and nothing else. It is important to understand that reactance *only* has a 90-degree phase shift, and no other value of phase shift. Reactance, for example, cannot have a 45-degree phase shift under any conditions. (It is only impedance, discussed in the next chapter, that can have *any* value of phase shift between +/–90 degrees.)

[2] It is not wrong to describe resistance as impedance, but more properly resistance is only the "real" part of impedance. It is not wrong to describe reactance as impedance, but more properly it is only the "imaginary" (shifted 90 degrees) component of impedance.

Reactance is further divided into two components:

- *Capacitive reactance* (designated by the symbol X_C) is inversely related to the size of the capacitor and to frequency. That is, larger capacitors have lower reactance at the same frequency, and a given capacitor has a lower reactance as frequency increases. Voltage *lags* current (current leads voltage) through a capacitor, so there is a (negative) 90-degree phase shift associated with capacitive reactance.
- *Inductive reactance* (designated by the symbol X_L) is directly related to the size of the inductor and to frequency. That is, larger inductors have higher reactance at the same frequency, and a given inductor has a higher reactance as frequency increases. Voltage *leads* current (current lags voltage) through an inductor, so there is a (positive) 90-degree phase shift associated with inductive reactance.

It is sometimes helpful to note that capacitive and inductive reactance are almost exactly opposite in their characteristics and effects. It's as if they were at extreme opposite ends of the same scale. Resistance, then, is a very special case exactly between these two extremes.

CAPACITIVE REACTANCE

It would follow from the preceding that the measure of capacitive reactance, for example, might be given by the formula $X_C = -1/($frequency x C$)$. (The minus sign means a minus 90-degree phase shift. It is a consequence of the "imaginary operator," something we introduce in the next chapter.) To use this formula we need to know what measure of frequency we need to use. Recall from the discussion in Chapter 1 there are several ways we can measure, or describe, frequency. The measure used here is the angular frequency, ω (which is $2\pi f$; refer back to Equation 1-2). So the definition of capacitive reactance is as shown in Equation 6-1:

$$X_C = -1/\omega C \text{ ohms} \qquad [6\text{-}1]$$

For example, suppose we have a .01 µF (10^{-8}) capacitor at an angular frequency of $\omega = 10^6$. (Remember an angular frequency of 10^6 would equate to a cyclical frequency of $10^6/2\pi = 159{,}155$ kHz). The reactance associated with this capacitor at this frequency would be:

$$X = -1/(10^6 \times 10^{-8}) = -100 \; \Omega$$

(The minus sign means that the voltage phase shift is –90 degrees).

This same capacitor at a frequency 10 times higher ($\omega = 10^7$) would have a reactance of:

$$X = -1/(10^7 \times 10^{-8}) = -10 \; \Omega$$

demonstrating the importance of the frequency of the circuit.

A larger capacitor, say .1 uF (10^{-7}) at a frequency $\omega = 10^6$ would have a reactance of

$$X = -1/(10^6 \times 10^{-7}) = -10 \; \Omega$$

These examples illustrate the importance of both the value of the capacitor and the frequency in determining the reactance (in ohms) the device presents to the circuit.

The relationship between capacitive reactance and frequency is nonlinear. That is, a graph of the relationship would be a curved line (see Figure 6-1). Even though we calculate capacitive reactance as a negative number indicating phase shift, we graph it with a positive sign. The reason for that will become clearer in the next chapter, when we show that $Z = \sqrt{X^2}$. Since frequencies of possible interest cover a very wide range (from a few Hz to hundreds of MHz and more), we usually graph the reactive relationship on "log–log" axes. The scales for both the horizontal and vertical axes are logarithmic (common log, to the base 10.) See Appendix C for more information about logarithms. When graphed this way (Figure 6-2) the capacitive reactance relationship is a straight line. Notice in what direction the graphs move as capacitance changes.

INDUCTIVE REACTANCE

Inductive reactance is given by the relationship shown in Equation 6-2:

$$XL = +\omega L \qquad\qquad [6\text{-}2]$$

where XL is inductive reactance in ohms
  ω is angular frequency and
  L is inductance in henrys
  The + sign (usually understood and not used) means a +90-degree phase shift.

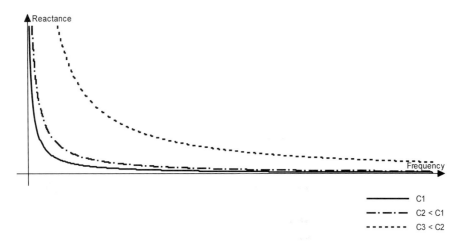

Figure 6-1 Capacitive reactance graphed on linear scales.

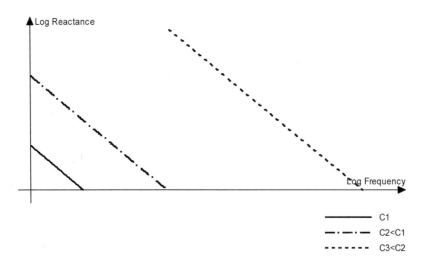

Figure 6-2 Capacitive reactance graphed on logarithmic scales.

For example, consider a component lead with 10 nH (10^{-8}) of inductance at a frequency $\omega = 10^6$. X_L would be $(10^6)(10^{-8}) = 10$ mΩ. But the same lead at a frequency of $\omega = 10^8$ would present a reactance of $X_L = (10^8)(10^{-8}) = 1.0$ Ω.

Inductive reactance is a linear relationship. That means that it would graph as a straight line whether linear or logarithmic scales were used (see Figure 6-3). We usually use logarithmic scales to graph inductive reactance because the range of frequencies is so large.

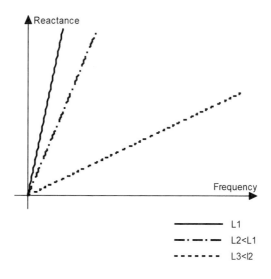

Figure 6-3 Inductive reactance graphed on logarithmic scales.

OHM'S LAW FOR REACTANCE

Just as Ohm's Law applies to resistance, Ohm's Law also applies to reactance (see Equation 6-3):

$$V = iX \qquad [6\text{-}3]$$

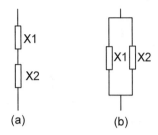

Figure 6-4 Series (a) and parallel (b) combinations of reactance.

Reactances combine just as resistances do (Figure 6-4). The equivalent reactance, Xeq, for two reactances in series is shown in Equation 6-4:

$$X_{eq} = X_1 + X_2 \qquad [6\text{-}4]$$

Reactances in parallel combine using the normal parallel combination formula as shown in Equation 6-5:

$$X_{eq} = \frac{1}{\dfrac{1}{X_1} + \dfrac{1}{X_2} + \cdots + \dfrac{1}{X_n}} \qquad [6\text{-}5]$$

From here we can see how the formula for an equivalent capacitance of two or more capacitors in series is derived (see Equation 6-6):

$$X_{eq} = X_1 + X_2 + \ldots + X_n$$

$$\frac{1}{\omega C_{eq}} = \frac{1}{\omega C_1} + \frac{1}{\omega C_2} + \ldots + \frac{1}{\omega C_n}$$

$$C_{eq} = \frac{1}{\dfrac{1}{C_1} + \dfrac{1}{C_2} + \ldots + \dfrac{1}{C_n}} \qquad [6\text{-}6]$$

When we place two or more capacitors in parallel the result is shown in Equation 6-7:

$$Xeq = \frac{1}{\frac{1}{X1} + \frac{1}{X2} + \cdots + \frac{1}{Xn}}$$

$$\frac{1}{\omega Ceq} = \frac{1}{\omega C1 + \omega C2 + \ldots + \omega Cn}$$

$$Ceq = C1 + C2 + \ldots + Cn \qquad [6\text{-}7]$$

Similarly, two or more series inductors combine as shown in Equation 6-8:

$$Xeq = X1 + X2 + \ldots + Xn$$
$$\omega Leq = \omega L1 + \omega L2 + \ldots + \omega Ln$$
$$Leq = L1 + L2 + \ldots + Ln \qquad [6\text{-}8]$$

Two or more inductors in parallel combine as shown in Equation 6-9:

$$Xeq = \frac{1}{\frac{1}{X1} + \frac{1}{X2} + \ldots + \frac{1}{Xn}}$$

$$\omega Leq = \frac{1}{\frac{1}{\omega L1} + \frac{1}{\omega L2} + \ldots + \frac{1}{\omega Ln}}$$

$$Leq = \frac{1}{\frac{1}{L1} + \frac{1}{L2} + \ldots + \frac{1}{Ln}} \qquad [6\text{-}9]$$

Series LC Combinations

It gets more interesting if we combine inductive and capacitive reactances. Suppose we have a capacitor and an inductor connected in series (Figure 6-5). The equivalent reactance for this circuit is shown in Equation 6-10:

$$\begin{aligned} X_{total} &= X1 + X2, \text{ or} \\ &= XL + XC \\ &= \omega L - 1/\omega C \\ &= (\omega^2 LC - 1)/\omega C \end{aligned} \qquad [6\text{-}10]$$

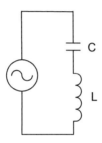

Figure 6-5 Series LC circuit.

Assume we have an AC current flowing through this circuit. Remember that current must be constant everywhere within a loop. This current (by Ohm's Law) generates a voltage across the capacitor, a voltage across the inductor, and a voltage across the pair. Ohm's Law applies for each case. Now the voltage across the capacitor *lags* the current by 90 degrees. The voltage across the inductor *leads* the current by 90 degrees. Therefore the voltage across the capacitor must *lag* the voltage across the inductor *by 180 degrees*.

This is a very important point. Since there are only 360 degrees in a cycle, that means the voltages across the capacitor and inductor are *exactly* opposite in phase. And they are *always* so. Always! Let's look at a couple of examples. Suppose we let:

$$\begin{aligned} C &= .1 \text{ uF} = 10^{-7} \\ L &= 4 \text{ uH} = 4 \times 10^{-6} \end{aligned}$$

Let current = 5 mA at an angular frequency of $2\pi f = 10^6$

$$XC = -1/\omega C = -1/(10^6)(10^{-7}) = -10 \; \Omega$$
$$XL = \omega L = (10^6)(4 \times 10^{-6}) = 4 \; \Omega$$
$$Xtotal = XC + XL = -6 \; \Omega$$

(Remember, the minus sign means that the voltage lags current.)

We should get the same total reactance if we calculate it from Equation 6-10:

$$Xtotal = [(10^6)(10^6)(10^{-7})(4 \times 10^{-6}) - 1]/(10^{-7})(10^6)$$
$$= -.6/.1 = -6 \; \Omega$$

The three voltage waveforms, then, would be:

$$VC = iXC = 5(10^{-3})(-10) \text{ with } -90\text{-degree phase shift}$$

$$VL = iXL = 5(10^{-3})(4) \text{ with } +90\text{-degree phase shift, and}$$

$$Vtotal = iXtotal = 5(10^{-3})(-6) \text{ with } -90\text{-degree phase shift}$$

The three voltage waveforms are graphed in Figure 6-6. There are several things to note from this example. First, the voltages across the capacitor and the inductor are exactly opposite in phase. Second, the total voltage across the circuit is the algebraic sum of the two voltages. (It could be determined graphically by simply adding the two individual curves.) Third, since the reactance of the capacitor is larger than the reactance of the inductor, the voltage across the capacitor is (naturally) larger than the voltage across the inductor. Consequently, the voltage phase shift of the total voltage is the same as the voltage phase shift across the capacitor. That is, for the total circuit the phase shift is –90 degrees, the voltage lags the current.

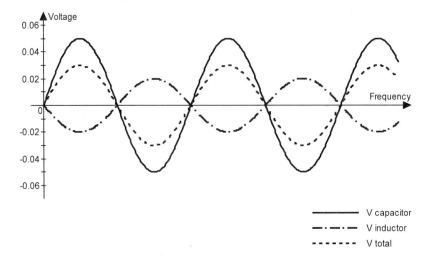

Figure 6-6 Plots of the three voltages at $\omega = 10^6$.

It is important to recognize that since the phase shift of the pair of components is –90 degrees, this circuit looks more like a capacitor than an inductor. This makes sense since the capacitive reactance is more than twice the magnitude of the inductive reactance. We call this circuit *capacitive* (at *this* frequency).

Now let's make just one change in this example. Let's change the frequency from $\omega = 10^6$ to $\omega = 2 \times 10^6$. Intuitively we expect that the capacitive reactance will go down and the inductive reactance will go up. We calculate:

$$XC = -1/\omega C = -1/(2 \times 10^6)(10^{-7}) = -5 \text{ } \Omega$$
$$XL = \omega L = (2 \times 10^6)(4 \times 10^{-6}) = 8 \text{ } \Omega$$
$$Xtotal = XC + XL = 3 \text{ } \Omega$$

Now the total reactance phase shift is +90 degrees, which means the total voltage is exactly in phase with the voltage across the inductor. The voltage curves now are shown in Figure 6-7. This figure looks very similar to Figure 6-6, except the voltages across the two components are different (they are still 180 degrees different in phase from each other), and the total voltage is now in phase with the voltage across the inductor.

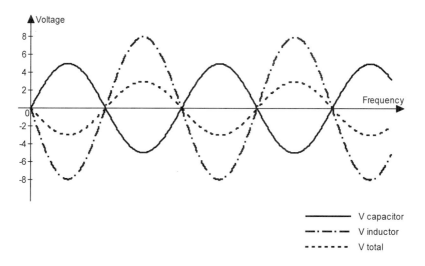

Figure 6-7 Plots of the three voltages at ω = 2 x 10⁶.

It is important to recognize that since the phase shift of the pair of components is +90 degrees, this circuit looks more like an inductor than a capacitor. This makes sense since the inductive reactance is now larger than the capacitive reactance (totally because the frequency, and *nothing* else, has changed). We now call this circuit *inductive* (at *this* frequency).

Note that this circuit has gone from a capacitive circuit to an inductive circuit without changing any component values at all. The change is solely the result of a change in frequency. The change in frequency wasn't really very much, just from ω = 10⁶ to ω = 2 x 10⁶. This example could be that of a normal bypass capacitor with lead inductance. Figure 6-5 could describe such a bypass capacitor exactly. (We introduce the impact of series resistance, ESR, in Chapter 15.) The bypass capacitor changes from a capacitive device to an inductive device simply as a result of the frequency increasing beyond a certain value.

PARALLEL LC CIRCUITS

Figure 6-8 illustrates a parallel LC circuit formed with the same components used in the series LC example. Now, however, the total reactance presented by the parallel combination is shown in Equation 6-11:

$$X_{total} = 1 / [(1/XC) + (1/XL)]$$
$$= 1/[-\omega C + 1/\omega L]$$
$$= \omega L/(1 - \omega^2 LC) \qquad [6\text{-}11]$$

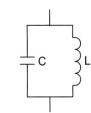

Figure 6-8 Parallel LC circuit.

At a frequency of $\omega = 10^6$ the values work out to be:

$$XC = -1/\omega C = -1/(10^6)(10^{-7}) = -10 \ \Omega$$
$$XL = \omega L = (10^6)(4 \times 10^{-6}) = 4 \ \Omega$$
$$X_{total} = (10^6)(4 \times 10^{-6})/[1 - (10^6)(10^6)(4 \times 10^{-6})(10^{-7})]$$
$$= 4/.6 = 6.66 \ \Omega$$

(Note: Xtotal does not equal XC + XL in this case, as you might expect, because XC and XL are connected in parallel, not series.)

If the voltage across the parallel combination is held at 5 volts with zero phase shift, we can calculate that the current through the components will be:

$$iC = 5/(-10) = -.5 \text{ amps}$$
$$iL = 5/4 = +1.25 \text{ amps}$$
$$i_{total} = 5/6.666 = +.75 \text{ amps.}$$

These waveforms are graphed in Figure 6-9. The minus (−) signs mean that the current through the capacitor leads the voltage (the voltage lags the current) by 90 degrees. The current through the inductor lags the voltage (the voltage leads the current) by 90 degrees. Therefore, the current through the capacitor leads the current through the inductor by 180 degrees. This relationship (between the parallel capacitor and inductor) cannot change; their currents are always out of phase by this 180 degrees.

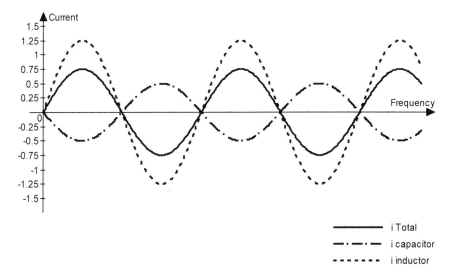

Figure 6-9 Parallel currents for $\omega = 10^6$.

The total current is the sum of the two currents, consistent with Kirchhoff's 1st Law (the current into a node must equal the current going out of a node). Since the sign of the total reactance is positive, we say this parallel combination is *inductive* at this frequency. That is, most of the current is flowing through the inductor, with only a small amount through the capacitor. If we extend this toward the limit at very low frequencies, the capacitor would have almost no effect on the circuit at all.

As we did before, let's just change one thing, frequency, to $\omega = 2 \times 10^6$. Now we find:

$$XC = -1/\omega C = -1/(2 \times 10^6)(10^{-7}) = -5 \; \Omega$$
$$XL = \omega L = (2 \times 10^6)(4 \times 10^{-6}) = 8 \; \Omega$$
$$Xtotal = (2 \times 10^6)(4 \times 10^{-6})/[1 - (2 \times 10^6)(2 \times 10^6)(4 \times 10^{-6})(10^{-7})]$$
$$= 8/(-.6) = -13.3333 \; \Omega$$

If the voltage across the parallel combination is held at 5 volts with zero phase shift, we can calculate that the current through the components will be the following:

$$iC = 5/(-5) = -1 \text{ amp}$$
$$iL = 5/8 = .625 \text{ amps}$$
$$itotal = 5/(-13.333) = -.375 \text{ amps}$$

These waveforms are graphed in Figure 6-10. As before, the minus (–) signs mean that the current through the capacitor leads the voltage (the voltage lags the current) by 90 degrees. The current through the inductor lags the voltage (the voltage leads the current) by 90 degrees. Therefore, the current through the capacitor leads the current through the inductor by 180 degrees. This relationship (between the parallel capacitor and inductor) cannot change; their currents are always out of phase by this 180 degrees.

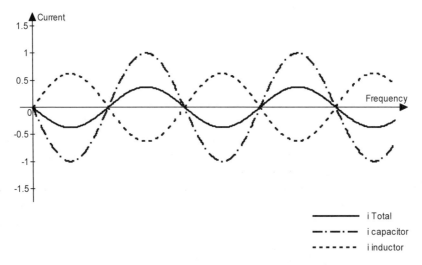

Figure 6-10 Parallel currents for $\omega = 2 \times 10^6$.

Note also as before, the total current is the sum of the two currents, consistent with Kirchhoff's 1st Law. But this time, since the sign of the total reactance is negative, we say this parallel combination is *capacitive* at this frequency. That is, most of the current is flowing through the capacitor, with only a small amount through the inductor. If we extend this toward the limit at very high frequencies, the inductor would have almost no effect on the circuit at all, and all the current would flow through the capacitor.

RESONANCE

When we plot the impedance of a circuit against frequency, we often refer to the graph as an impedance plot. Such a graph is usually plotted against log–log axes. Figure 6-11 shows impedance plots of the serial and parallel LC circuits described in Figures 6-5 and 6-8, respectively. Even though the reactance of a capacitive circuit is negative, we plot the impedance as a positive number on the axes. More properly, we say we plot the *absolute value* (the value without regard to sign).[3]

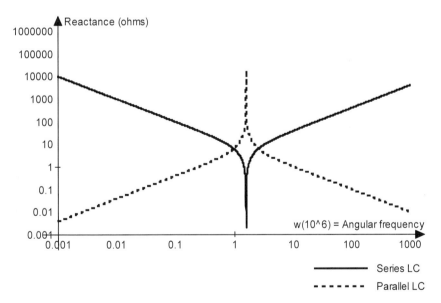

Figure 6-11 Impedance plots for the series and parallel circuits of Figures 6-5 and 6-8.

---

[3]When you stop to think about it, you can't have negative impedance. That would imply that a positive voltage, for example, would result in a negative current, a current flowing in the wrong direction. We will see in the next chapter how we get a positive impedance value from a negative reactance.

We often say that the capacitive portion of an impedance curve is "downward sloping to the right." This is true for both of these curves. On any impedance curve, for any area in which the impedance is falling as the frequency is increasing, the circuit will look capacitive (the current will lead the voltage through the circuit at those frequencies).

Each of the curves in Figure 6-11 exhibits an interesting characteristic at a particular frequency. The frequency is the same in each curve, about 250 kHz, where $\omega = 1.58 \times 10^6$. To see what is special about this point, look at Equations 6-10 and 6-11. Consider the term $\omega^2 LC$. If this term exactly equals 1.0, then the numerator of the series reactance term (Equation 6-10) goes to exactly zero. That means there is zero reactance (impedance) through the series LC circuit. (Not just a small reactance; absolutely *zero* reactance.[4]) The frequency at which this term goes to zero is found by setting (in Equation 6-12):

$$\omega^2 LC = 1 \qquad [6\text{-}12]$$
$$\omega^2 = 1/LC = 1/(4 \times 10^{-6})(10^{-7}) = 2.5 \times 10^{12}$$
$$\omega = 1.58 \times 10^6 = 2\pi f$$
$$f = 1.58 \times 10^6 / 2\pi \cong 251{,}646 \text{ kHz}$$

This does *not* mean that there is no voltage across each component. There is current flowing, and by Kirchhoff's 2nd Law (see Chapter 5), the current is constant everywhere in the loop. The reactance through the capacitor is not zero, nor is the reactance through the inductor. Since there is reactance through each component, then the current will introduce a voltage across each component, even at this particular frequency. What is special about this particular frequency is that the reactance of each component is *exactly* equal at this frequency. That means a common current will create *exactly* equal voltages across each component. And since the voltages across a capacitor and an inductor are exactly 180 degrees out of phase, their *sum* is exactly zero and they exactly cancel. That is why there is no voltage that can exist across the series pair *at this particular frequency*.

At this same frequency, the denominator of the parallel LC reactance term (Equation 6-11) also goes to zero. At that point, the parallel reactance goes to infinity. (Not just very high; infinitely high.) It's as if there is an open circuit at that particular frequency.

---

[4]From a practical standpoint there must be some residual (parasitic) resistance in any circuit. In this chapter, however, we are considering "ideal" capacitors and inductors, those without any resistance.

That does not mean that there is no current flowing through the components. As can be seen from Figure 6-8, the voltage across the capacitor is exactly the same as the voltage across the inductor. At this particular frequency, the reactance of the capacitor is exactly the same as the reactance of the inductor. Therefore, the current through the capacitor must be exactly the same as the current through the inductor. But since the phase of the currents differs by exactly 180 degrees, the currents are equal and opposite. Therefore, the current into the node from one component is exactly the same as the current out of the node toward the other component.

By Kirchhoff's 1st Law the total current into the node must equal the total current out of the node. Under this condition that means there is zero current that can enter the node from the voltage source. Since there is zero current from the voltage source, the voltage source must be looking into an infinitely high impedance.

This very special frequency (defined by Equation 6-12), where

$$\omega^2 LC = 1$$
$$\omega^2 = 1/LC$$
$$\omega = 1/\sqrt{LC} = 2\pi f \text{ or}$$
$$f = 1/(2\pi \sqrt{LC}) \qquad [6\text{-}13]$$

is called the *resonant* frequency of the circuit (see Equation 6-13). Resonance is a very important concept. One place where its use might be obvious is in a tuning circuit. If we placed a series LC circuit between two stages of an amplifier, we can envision that the only frequency that would pass through from one stage to another is the resonant frequency. All other frequencies would (tend to be) blocked by the LC circuit. If we placed a parallel LC circuit to ground between two stages of an amplifier, the resonant frequency would be passed on to the next stage. All other frequencies would be shunted toward ground. Of course, in real life, circuits are a little more complicated than this, but this principle is the one on which most tuning circuits are based.

Consider the case of a typical bypass capacitor, say one with a value of .01 µF and a lead inductance of 5 nH. This would look to the circuit like a series LC circuit whose reactance would be given by the expression:

$$X = X_L + X_C$$
$$= \omega(5)(10^{-9}) - 1/\omega(10^{-8})$$
$$= [5\omega^2(10^{-9})(10^{-8}) - 1]/\omega(10^{-8})$$

Since this capacitor has some associated inductance, there must be a resonant frequency. The resonant frequency of this capacitor is

$$f = 1/(2\pi\sqrt{LC}) \cong 22.5 \text{ MHz}$$
(22,507,926.92 Hz to be more exact!)

Figure 6-12 is a plot of this impedance function for this bypass capacitor. Included also is a plot of an ideal .01 capacitor and an ideal 5 nH inductor. Note that the bypass capacitor plot is capacitive (what we refer to as downward sloping to the right) until the resonant frequency, then becomes upward sloping (inductive) past that point. Beyond 25.5 MHz the capacitor actually starts looking like an inductor. That is the point beyond which the inductive reactance begins to exceed the capacitive reactance.

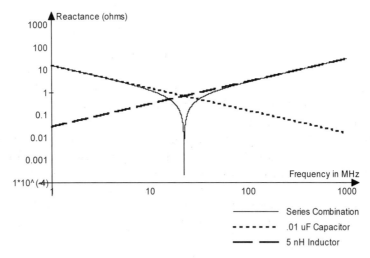

Figure 6-12 Impedance plot of the bypass capacitor and each component individually.

At frequencies well below the resonant frequency, the capacitive reactance dominates; the inductive reactance is so small that it is negligible. At frequencies well above the resonant frequency the inductive reactance dominates and the capacitive reactance is so small that it is negligible. But at frequencies near resonance, the two interact in such a way that the combined effect is lower impedance than either component taken separately.

There are two important points to recognize in this example. The first is that the phase shift caused by this bypass capacitor is either –90 degrees (capacitive) or +90 degrees (inductive). It is never anything else (except that the phase shift is *exactly* zero at *exactly* the resonant frequency). We sometimes think of this as "shooting" rapidly from –90 degrees, through zero, to +90 degrees. In fact it does not progress somehow from one extreme to the other. There is a discontinuity at the resonant frequency. It is –90 degrees below the resonant frequency, no matter how close to the resonant frequency you want to get. It is +90 degrees above the resonant frequency, no matter how close to the resonant frequency you want to get. It is exactly zero at exactly the resonant frequency. It is *never* anything else. (Remember, we are assuming no resistance at all.)

There are those who might say that above the resonant frequency this bypass capacitor (a) looks like an inductor and therefore (b) is useless in the circuit. Comment (a) is true, but (b) does not follow logically from it. First note that even (for a little ways) above the resonant frequency the impedance of the real device is still *below* that of the ideal capacitor, because of the interaction with the inductance. This effect becomes more pronounced in circuits that have many capacitors, as we will see in Chapter 15. Second, it is not necessarily the bypass capacitor effect that is important in our circuits; it is often the net impedance offered by the device. Even in the inductive region the net impedance can be very small, and therefore the device is providing a "service" to the circuit. Be careful of people who make absolute statements about components either having or not having an effect at certain frequencies. It is not necessarily that simple.

POLES AND ZEROS

Figure 6-13 illustrates a case with two series LC circuits in parallel. We will look at them for now as just that; but later we will see that this is exactly the case that we have when C1 and C2 are bypass capacitors and L1 and L2 represent the stray inductance contributed by leads, pads, vias, and traces.

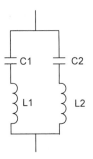

Figure 6-13 Schematic of two parallel capacitors, each with associated lead and mounting inductance.

Let these components have the following values:

$$L1 = 1.0 \text{ nH} = 10^{-9}$$
$$L2 = .1 \text{ nH} = 10^{-10}$$
$$C1 = .1 \text{ uF} = 10^{-7}$$
$$C2 = .01 \text{ uF} = 10^{-8}$$

Remember that the resonant frequency of a series LC pair is given by Equation 6-13:

$$f = 1/(2\pi \sqrt{LC}\,)$$

Using this formula we can find that these pairs will resonate at about 16 MHz and 160 MHz, respectively. Figure 6-14 plots the impedance function of each of these LC circuits. At lower frequencies each circuit looks more like a capacitor than an inductor. But above each resonance, along the upward-sloping part of the reactance curve, each circuit looks more like an inductor than a capacitor. Over this rising portion of the curve the sign of the reactance would be positive.

Now here comes some interesting stuff. Look on the graph at about where the frequency would be 50 MHz. The left-hand curve (Pair 1) is inductive and its reactance has a positive sign here. The downward-sloping curve (Pair 2) is capacitive and its reactance here is negative. The curves are plotted for each one separately, as if the other one weren't there. But in the circuit, of course, they *are* both there and their effects combine. One hint about the way they are going to combine is the fact that at (approximately) 50 MHz their reactances are equal and opposite. This implies another resonance point. But these *pairs* are not in series, they are in *parallel*. So, if there is another resonance point, it will be a parallel resonance point and the effect must be infinite reactance.

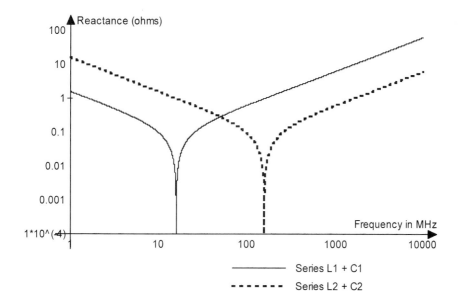

Figure 6-14 Impedance function of each capacitor separately.

Another way to view this is to recognize that the four components do form a parallel circuit of Ls and Cs. At some frequencies one side will look mostly like an inductor and the other side will look mostly like a capacitor. Therefore, it makes sense that there would be a parallel resonance formed by this set somewhere in that frequency range.

And indeed there is! Figure 6-15 plots the combined effect of all the components making up this circuit. The parallel resonance point between the two series resonance points is very clear. Not coincidentally, it is at the precise point where the two individual curves cross.

The series and parallel resonance points in Figure 6-15 are very sharp because we have included no resistance effects in the analysis. In real life there is always some resistance around, so results on a PCB would normally not be as sharp as is shown here, which is fortunate for all of us. The impact of resistance on LC circuits will be explored in the next chapter and in Chapter 15.

The series resonance points, where the impedance goes to zero, are called (appropriately enough) *zeros*. The parallel resonance points are called *poles.* It seems intuitive that they are called that because they stick up on the frequency plots. In a 3D reactance or impedance plot the poles and zeros can look just like they are holding up or tying down a sheet of canvas for a circus tent.

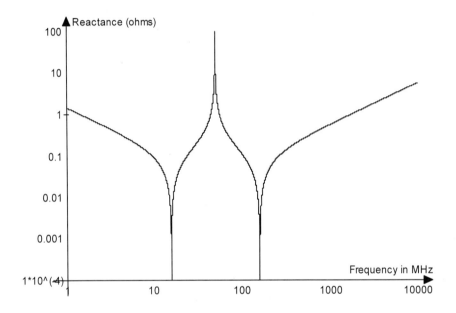

Figure 6-15 Combined impedance of L1 and C1 in Parallel with L2 and C2

There is another reason for looking at the concept of poles and zeros. It is true that:

1. If there are two zeros, there will be a pole in between.
2. If there are two poles, there will be a zero in between.
3. Anytime there is a pole, there will also be a zero.
4. Anytime there is a zero, there will also be a pole.
5. It is possible that some the poles and zeros occur at the same frequencies.

This can be helpful in understanding how circuits will react and a where certain unexpected results might be coming from.

Now, perhaps you have already said, "Well, wait a minute. If I have a single capacitor with a zero at infinity, where is the pole?" The answer is at DC (zero frequency.) An inductor, with a pole at infinitely high frequencies, has a zero at DC. A single series combination LC circuit has a zero at its resonant frequency. It has poles at DC and at infinite frequency. A parallel LC circuit with a pole at its resonant frequency has zeros at both DC and at infinite frequency. Poles and zeros exist in pairs and one will always be accompanied by the other.

Later we will see that capacitors almost always have some associated inductance with them, so every capacitor probably has a series resonant point (a zero). One could wonder if we have 200 bypass capacitors on our board, and therefore 200 zeros, will we also have 200 parallel resonance points, or poles? The answer is yes, this is true. This can have a serious impact on bypass capacitor and power supply strategy, and can be an important design consideration. We will talk more about this later, particularly in Chapter 15. One clue why we don't have more problems than we already do, however, is because there are a lot of effective resistance paths that help attenuate the sharp peaks. The real world situation is not quite as bad, therefore, as is suggested by Figure 6-15.

SUSCEPTANCE

In the interest of thoroughness I introduce here the term *susceptance*. Susceptance is the inverse of reactance, just as conductance is the inverse of resistance (see Chapter 5). Susceptance is denoted by the symbol B and is defined as:

$$B = 1/X$$

The phase shift associated with susceptance is the same as the phase shift associated with reactance. That is, if the capacitive (voltage) phase shift is −90 degrees, then capacitive susceptance will also have a (voltage) phase shift of −90 degrees.

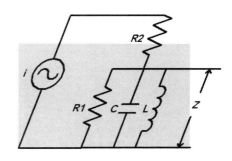

7 IMPEDANCE AND PHASE SHIFT

IMPEDANCE

*M*ost people find that analyses regarding resistance are fairly easy. Analyses regarding capacitive reactance aren't too difficult, because capacitive effects seem to be somewhat intuitive to a lot of people. The effects of inductors are much less intuitive, and people tend to have more problems with them. And then there is impedance.

Table 7-1 Summary of the Effects of the Primary Passive Components

Component	"Impedance" to Current	Measured in	Voltage Phase	Frequency Dependent?	Impedance Is High if
R	Resistive (R)	ohms	0	No	R ⇒ high
L	Inductive reactance (X)	ohms	+90	Yes	L ⇒ high f ⇒ low
C	Capacitive reactance (X)	ohms	−90	Yes	C ⇒ low f ⇒ high

Consider the simple RC circuit shown in Figure 7-1, driven by an AC constant current source.[1] If the value of the capacitor is very large, there will be very little voltage across it. The voltage across the RC pair will be almost identical to the

---

[1] A constant current source maintains a constant current and phase under all conditions. Constant current sources are harder to design than constant voltage sources, but they are not uncommon in circuits.

voltage across the resistor, and it will be in phase (i.e., zero phase shift) with respect to the current. If the value of the capacitor is very small, then the voltage across it will be much larger than the voltage across the resistor. The voltage across the RC pair will be almost equal to the voltage across the capacitor. Its voltage phase will be –90 degrees (90 degrees behind) the phase of the current.

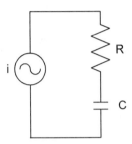

Figure 7-1 Simple RC circuit driven by a constant current source.

Therefore, the voltage across the RC pair ranges from 0-degrees phase shift for large values of capacitance to –90-degrees phase shift for small values of capacitance. There must, therefore, be a large number of values of capacitance for which the phase shift across the RC pair will be somewhere in between these values, say 45 degrees, 57 degrees, or 82 degrees, for example.

Resistors offer resistance to the flow of current. Resistance is really a very special case of impedance, that of zero phase shift. Capacitors and inductors resist the flow of current with reactance. Reactance is also a very special case of (exactly) 90-degree phase shift (either plus or minus). The general case of resistance to current flow is called *impedance* and it covers all cases of phase shift from +90 degrees to –90 degrees (including +90 degrees, –90 degrees and zero-degree phase shift). Since impedance is the general case, we can say that pure resistance is always a measure of impedance, but only in a very special case would impedance be purely resistive.

Formally, impedance (Z) is the combination of resistance (R) and reactance (X) of a circuit according to the expression shown in Equation 7-1:

$$Z = R + jX \qquad [7\text{-}1]$$

The symbol j stands for the square root of –1, $\sqrt{-1}$, which, of course, doesn't exist (see the Chapter Endnote on imaginary numbers at the end of this chapter and also see Appendix E). The impedance expression has what we call a real part (R) and an imaginary part (X). The practical distinction between these parts is that they are *orthogonal*, or perpendicular, to each other. That is, they are shifted from each other by 90 degrees. This is, in fact, the 90-degree phase shift between reactance and resistance.

When we look at impedance we usually picture it as being graphed on axes as shown in Figure 7-2. R is usually graphed on the horizontal (zero phase) axis, and reactance is graphed on the vertical (90-degree phase) axis. So inductive reactance (+90 degree) would be plotted on the vertical axis above the horizontal axis, and capacitive reactance (–90 degree) would be plotted on the vertical axis below the horizontal axis.

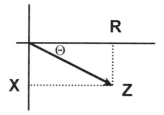

Figure 7-2 For any given R and X, impedance is the vector to their intersection and the phase shift is the angle between that vector and the horizontal axis.

Impedance is referred to as a vector, Z. A vector has magnitude and it has direction (angle from some reference, usually the origin, the intersection of the horizontal and vertical axes). Z can be found graphically by plotting the intersection of R and X (Figure 7-2). The magnitude of Z is the length of the vector from the origin to the intersection. The phase shift is the angle (Θ) formed between the vector and the horizontal axis. The magnitude of Z can also be solved for mathematically. It is the hypotenuse of the right triangle formed by R, X, and Z and is found by the Pythagorean theorem (Equation 7-2) we all learned in trigonometry.[2]

---

[2]Now we can see why we treat negative capacitive reactance as a positive impedance. If R = 0 and XC is negative, $Z = \sqrt{(-XC)^2} = XC$ (a positive value).

$$Z=\sqrt{R^2+X^2} \qquad [7\text{-}2]$$

Return now to Figure 7-1. If the capacitor is very large, X is very small, and Z is almost the same as R, with no phase shift. But if the capacitor is very small, then X is very large, Z is almost the same as X, and the phase shift is almost the same as that for a pure capacitor, –90 degrees.

Let's look now at a more normal case. Let's say R = 100 Ω and C = .02 µF. But wait a minute! We can't plot that yet because we haven't specified a frequency. We must know the frequency before we can calculate X. So let's let $\omega = 10^6$.

Now we can calculate X as follows:

$$-1/\omega C = -1/(10^6 \times .02 \times 10^{-6}) = -50 \text{ }\Omega$$

(Remember, the minus sign means a capacitive, or negative, phase shift.) So Z, now, is given by Equation 7-3:

$$Z = \sqrt{100^2 + (-50)^2} = 111.8 \text{ }\Omega \qquad [7\text{-}3]$$

The angle, Θ, is the resulting phase shift of the RC pair. Clearly Θ will be negative because the circuit must look at least a little capacitive (there is no inductance). We can measure the angle Θ graphically, or we can calculate Θ from trigonometry. The tangent of an angle is the ratio of the side opposite (X) over the side adjacent (R). So the phase angle is found by finding the angle whose tangent is X/R, or by the expression in Equation 7-4:

$$\Theta = \tan^{-1}(X/R) \qquad [7\text{-}4]$$

The expression $\tan^{-1}$ is called the arctangent, and almost all current spreadsheets and many handheld calculators calculate it (be careful to know whether your calculator or spreadsheet is working in degrees or radians). In our illustration, we use Equation 7-5 to find Θ:

$$\Theta = \tan^{-1}(-50/100) = -26.6 \text{ degrees.} \qquad [7\text{-}5]$$

Figure 7-3 shows the waveforms for the circuit in Figure 7-1 for the following values:

$$i = \sin(\Theta)$$
$$\omega = 10^6$$
$$R = 100\ \Ega$$
$$C = .02\ \mu F$$
$$XC = -50\ \Omega$$

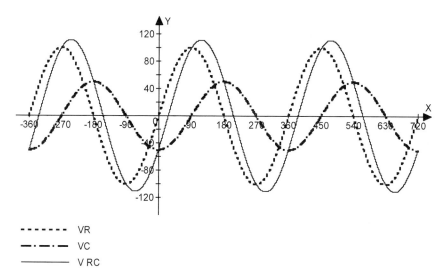

Figure 7-3 Voltage waveforms for the circuit in Figure 7-1.

The voltage across the resistor is:

$$VR = iR = 100\sin(\Theta)$$

It is in phase with the current and it is the curve in Figure 7-3 that goes through the origin (i.e., the value is 0 at 0 degrees).

The voltage across the capacitor is:

$$VC = iXC = -50\sin(\Theta) = 50\sin(\Theta - 90)$$

Here the minus sign means a phase shift of –90 degrees. So the voltage curve for VC is the curve whose value is 0 at 90 degrees (the 90-degree voltage lag). The voltage across the RC pair can be found two ways:

(a) We found from Equations 7-3 and 7-5 that $Z = 111.8 \, \Omega$ with a phase shift of –26.6 degrees. So we can use:

$$VRC = iZ = 111.8\sin(\Theta - 26.6)$$

This is the curve in Figure 7-3 that crosses the origin between the curves for VR and VC.

(b) We can also simply sum the two curves for VR and VC. That is:

$$VRC = VR + VC = 100\sin(\Theta) + 50\sin(\Theta - 90)$$

You can verify visually from Figure 7-3 that VRC is the arithmetic sum of the other two curves. If you go back to high school trigonometry, you can work this out through those formulas, but we won't do that here.

Now let's look at the same circuit, at the same frequency, but this time make both R and C a little smaller. Thus the resistance will drop but the capacitive reactance will increase. If we let:

$$R = 50 \, \Omega$$
$$C = .01 \, \mu F$$
$$\omega = 10^6$$
$$XC = -1/\omega C = -1/(10^6)(10^{-8}) = -100 \, \Omega$$

The voltage across the resistor is:

$$VR = Ri = 50i = 50\sin(\Theta)$$

The voltage across the capacitor is:

$$VC = XCi = -100i = 100\sin(\Theta - 90)$$

When we calculate the impedance we get:

$$Z = \sqrt{50^2 + 100^2}$$
$$= 111.8 \text{ }\Omega, \text{ and}$$
$$\Theta = \tan^{-1}(X/R)$$
$$= \tan^{-1}(-100/50)$$
$$= -63.4 \text{ degrees, and}$$
$$VRC = Zi = 111.8\sin(\Theta - 63.4)$$

These curves are drawn in Figure 7-4. Note that the relative phase of VR and VC *has not changed*, but their magnitudes have. The phase relationship between VR and VC cannot change. Voltage across a resistor is *always* in phase with the current and the voltage across a capacitor *always* lags current by exactly 90 degrees. As before, VR goes through the origin of the graph and VC crosses the horizontal axis at 90 degrees. Their magnitudes changed by such an amount that their combined impedance didn't change (I did that on purpose). Because the circuit is more capacitive now, the total voltage, VRC, has shifted further in phase. We can see this by comparing the impedance plot in Figure 7-5 with Figure 7-2.

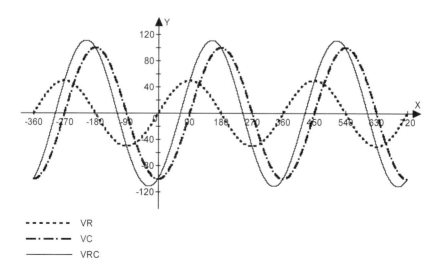

Figure 7-4 Voltage waveforms.

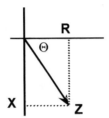

Figure 7-5 Impedance plot for Z = 111.8 Ω and Θ = –63.4 degrees.

EFFECT OF FREQUENCY

Reviewing what we have just done with our circuit in Figure 7-1, if:

$$R = 50 \, \Omega$$
$$C = .01 \, \mu F$$
$$\omega = 10^6$$
$$XC = -100 \, \Omega, \text{ then:}$$
$$Z = 111.8 \, \Omega, \text{ and}$$
$$\Theta = -64.3 \text{ degrees.}$$

This is what is plotted in Figure 7-5. Now let's change nothing except to let $\omega = 10^7$. With this change, $XC = -1/\omega C = -10 \, \Omega$. We have changed no components, only frequency. Now:

$$Z = \sqrt{50^2 + 10^2}$$
$$= 51.0 \, \Omega, \text{ and}$$
$$\Theta = \tan^{-1}(X/C) = \tan^{-1}(-10/50) = -11.3 \text{ degrees.}$$

This impedance is shown graphically in Figure 7-6, illustrating the truly dramatic effect that frequency has on impedance. All other things being equal (i.e., the Rs and Cs don't change), impedance and phase shift will change dramatically with frequency.

Even if we have a relatively low clock frequency associated with our circuit, if that clock signal has a fast rise time there will be some high-frequency harmonic signals through our circuits. In fact, there will be several transient harmonics at several different frequencies. Each of these harmonics will see our circuit differently because the impedance that our circuits present to a signal, and the phase shift it causes with our signal, is dependent on the frequency of the *particular* harmonic. Thus, if we are not careful, the string of harmonics (say representing a square wave) we inject into a circuit may look quite different when they come out of the other end of the circuit. One name for this difference in signal shape is *distortion*. This is what can make high-speed signal analysis so difficult, and why we have to design our circuits so they present as little distortion to our signal waveforms as possible.

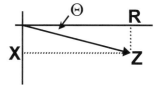

Figure 7-6 Impedance plot for Z = 51 Ω and Θ = –11.3 degrees.

It is beyond the scope of this book to cover how to make impedance measurements and calculations. The purpose instead is to give you an appreciation of what good, experienced engineers are faced with when they design fast, complex circuits. The rest of this chapter covers some simple concepts, such as RC filters, how impedances combine in circuits, and voltage waveforms in RLC circuits.

ANOTHER RC EXAMPLE

The illustration in Figure 7-1 used a constant current source to drive the RC circuit. Let's do the same analysis with the same circuit, but this time with a constant voltage source as shown in Figure 7-7.

Let:

$$V = 100\sin(\Theta)$$
$$\omega = 10^6$$
$$R = 100 \; \Omega$$

$$C = .02 \, \mu F$$
$$XC = -50 \, \Omega$$

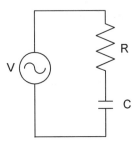

Figure 7-7 Simple RC circuit driven with a constant voltage source.

We have already calculated (Equations 7-3 and 7-5) Z as 111.8 Ω with a voltage phase shift, Θ, equal to –26.6 degrees. So current, by Ohm's Law, is:

$$i = V/Z = 100/111.8 = .894 \text{ amps with a phase of } +26.6$$
$$(\text{capacitive, current leading voltage}), \text{ or}$$
$$i = .894\sin(\Theta + 26.6)$$

The voltages across R and C, then, are:

$$VR = Ri = (100)(.894)\sin(\Theta + 26.6)$$
$$= 89.4\sin(\Theta + 26.6), \text{ and}$$
$$VC = XCi = (50)(.894)\sin(\Theta - 63.4)$$
$$= 44.7\sin(\Theta - 63.4)$$

(that is, VC lags both the current and the resistive voltage by 90 degrees).

It is tempting to look at this and say, therefore, that the total maximum voltage can be 89.4 + 44.7 = 134.1 volts. Don't make this mistake. We can't just total these coefficients. We must first adjust them for the phase and *then* add them. The total voltage is graphed in Figure 7-8 and is found by:

$$V_{total} = 100\sin(\Theta) = 89.4\sin(\Theta + 26.6) + 44.7\sin(\Theta - 63.4)$$

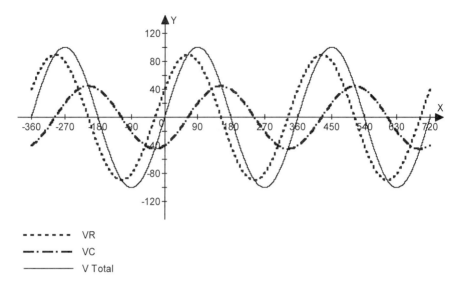

Figure 7-8 Voltage waveforms for this new example.

Table 7-2 in the notes at the end of this chapter shows the calculations for these terms for various angles between 0 and 90 degrees, showing that the calculations do, in fact, work.

CLASSIC RC FILTER

Remember our discussion about voltage dividers (Chapter 5)? Let's look at a classic RC filter circuit (Figure 7-9) from the standpoint of a voltage divider. This is called a *filter* circuit because we are going to filter out (shunt to ground) the higher frequencies and pass through only the lower frequencies. You have probably seen many of these circuits at the power input section of most PCBs.

Just as the voltage ratio of a resistive voltage divider is RL/RT, the voltage ratio here is:

$$Vout/Vin = ZC/ZRC$$

where ZC is the impedance of the capacitor and ZRC is the total impedance seen by Vin.

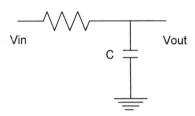

Figure 7-9 Classic RC filter commonly found in power supply circuits.

Since:
$$ZC = 1/\omega C$$
$$Z_{total} = R + jX = R + 1/j\omega C$$

The voltage ratio, then, becomes

$$\frac{V_{out}}{V_{in}} = \frac{Z_c}{Z_{total}} = \frac{\frac{1}{\omega C}}{\sqrt{R^2 + \left(\frac{-1}{\omega C}\right)^2}} = \frac{1}{\sqrt{1 + (\omega RC)^2}}$$

If we plot this equation against frequency (using a logarithmic scale), the curve looks like Figure 7-10. Lower frequencies are passed without attenuation, but higher frequencies get shunted to ground through the capacitor. The frequency where the voltage ratio falls to .707 is commonly called the *cutoff frequency* for the filter. That is the point at which $\omega = 1/RC$. We recognize $1/RC$ as the RC time constant we talked about in Chapter 4. Remember also that power is related to voltage squared. Noting that $(.707)^2$ is .5, we see that the RC filter cutoff point, $1/RC$, the RC time constant, also represents the point on the curve where *power* is reduced by 50%.

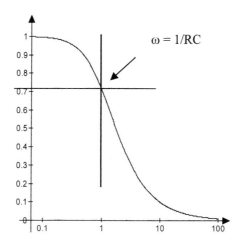

Figure 7-10 Frequency response of an RC filter.

COMBINING IMPEDANCES

We have already looked at how resistors, capacitors, inductors, and reactances combine in series and parallel. What about impedance? How do impedances combine in series and parallel?

Two impedances in series (Figure 7-11a) combine easily. If:

$$Z1 = R1 + jX1, \text{ and}$$
$$Z2 = R2 + jX2, \text{ then}$$
$$Ztotal = (R1 + R2) + j(X1 + X2)$$

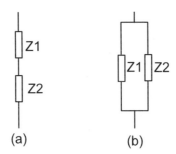

Figure 7-11 Combining impedances.

Parallel impedances pose a problem that is beyond the scope of this book. For practical purposes, the approach to this problem would be to reduce the impedances to their real and imaginary components, combine each of these components separately, and then recombine them into an impedance expression. The general solutions to complex impedance analyses can be very difficult. Since the reactive terms are functions of frequency, it is often easier to assume a *specific frequency*, and then work with the numerical values. Of course, if there are several frequencies of interest, this can be tedious, time consuming, and difficult to do without making errors. There comes a point where we stop doing this manually and start using higher order calculus and transforms and computerized programs to aid in the analyses.

RESONANCE AND Q

It is almost impossible to have an LC circuit without some resistance somewhere. Even if we don't add a resistor to achieve some effect, there is still parasitic resistance at other places in the circuit.

The impedance equation for a series RLC circuit can be developed as follows:

$$Z = \sqrt{R^2 + X^2} \text{ where}$$
$$R = \text{resistance and}$$
$$X = \text{reactance} = \omega L - 1/\omega C = (\omega^2 LC - 1)/\omega C$$

Figure 7-12b illustrates a series RLC circuit. We will let L = .1 µH and C = .1 µF. The resonant frequency for this LC pair is:

$$f = \frac{1}{2\pi\sqrt{LC}} = 316 \text{ kHz}$$

Figure 7-12a shows the impedance curve for this circuit as a function of frequency. It is centered on the resonant frequency. Three curves are shown, each for progressively higher values of R1. (The resistor R2 is there simply to establish a uniform maximum impedance for visual purposes.)

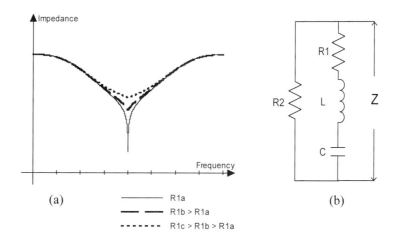

Figure 7-12 Series RLC circuit (a) and its impedance plot (b).

Recall from the discussion in Chapter 6 how sharp the resonance point could be at resonance. These curves illustrate how the frequency selectivity of an RLC series circuit depends heavily on the series resistance. Low values of series resistance (curve R1a in Figure 7-12a) result in a very sharp resonance point, or a very selective filter, while higher values of R1 (R1b and R1c in Figure 7-12a) create broader patterns as frequency varies. If we were tuning a radio, for example, we would likely want a very sharp selectivity. But we might want a broader pattern if we were building a band pass filter and wanted to pass a range of frequencies.

Figure 7-13b shows a parallel RLC circuit, this time driven by a current source. The plotted curves (Figure 7-13a) show the voltage across the RLC circuit as a function of frequency. The values of L and C are as before, with a resonant frequency of 316 kHz. If the parallel resistor, R1, is not there (or is a very high value), the voltage, at least in theory, can become almost infinitely high at resonance. This is because we are assuming a constant current driving an almost infinite parallel LC impedance at resonance. The resulting voltage, from Ohm's Law (V = iX) must also be almost infinite. But if there is some resistance in the circuit, the peak voltage will be limited by the value of that resistor times the current (Vmax = iR1). Thus, a lower value of parallel resistance, R1, will result in a lower peak voltage across the circuit. (As before, R2 is included in the circuit simply to improve the visual characteristics of the graph.)

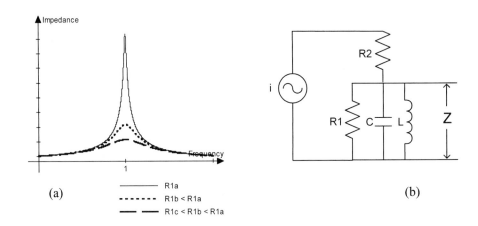

Figure 7-13 Impedance plot of a parallel RLC circuit.

The more sharply peaked curve in Figure 7-13 is for the highest resistance value for R1. The other curves are for progressively lower values of resistance. As before, the selectivity and effectiveness of this RLC filter depends heavily on the value of R1.

We sometimes talk about the Q of an RLC circuit. Q is called the quality factor of the circuit. For a series RLC circuit Q is defined as

$$Q = \omega_r L/R$$

where ω_r is the resonant frequency. As R gets very small, the bandwidth of the circuit gets very narrow and the Q of the circuit increases, sometimes to a quite high value.

The corresponding measure for a parallel RLC circuit is (perhaps not surprisingly)

$$Q = R/\omega_r L$$

again where ω_r is the resonant frequency. In the parallel circuit, as R increases, the bandwidth of the circuit gets narrower and the Q of the circuit increases.

On our circuit boards, bypass capacitors are really series RLC circuits. The C is, of course, the value of the capacitor. L is the inductance associated with the leads and terminals that are a part of the capacitor as well as any additional inductance associated with mounting the capacitor on the board. R is often referred to as equivalent series resistance (ESR) which is usually a part of the specification for a capacitor or capacitor family. Since bypass capacitors are really RLC circuits, they have a resonant frequency and a Q associated with them. If we connect two different capacitors together in parallel, there can be a *parallel* resonance between them, also characterized by a frequency and a Q. We will look more into these effects in Chapter 15.

Note: The design of frequency selective filter circuits (whether they are LC or RLC) can become very complicated very quickly. Filter design is a specialty by itself within the field of electronics, and is well beyond the scope of this book. All we can do here is give you a flavor for the most basic concepts behind filter design.

SERIES RLC CIRCUITS

A series RLC circuit can be really interesting to look at. Consider the circuit in Figure 7-14.

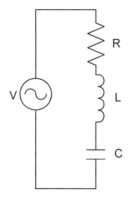

Figure 7-14 Series RLC circuit.

Let's let:

$$V_{in} = 10\sin(\Theta)$$
$$R = .1 \, \Omega$$
$$L = 5 \, nH$$
$$C = .1 \, \mu F$$
$$\omega = 50 \times 10^6$$

You might recognize that these are reasonable values for a bypass capacitor.

We pointed out earlier that the way to calculate the impedance of this RLC circuit is to first define its resistive and reactive parts. The resistive part is simply .1 Ω. Its reactive part is the series combination of the capacitive and inductive reactances, or

$$X = \omega L + (-1/\omega C)$$
$$= 50(10^6)(5)(10^{-9}) + (-1/50(10^6)(10^{-7}))$$
$$= .250 - .200 = (+).05 \, \Omega$$

Since the sign of X is positive, this circuit is slightly inductive at this frequency. Based on this, we can now calculate:

$$Z = \sqrt{.1^2 + .05^2} = .112 \, \Omega$$
$$\Theta = \tan^{-1}(.05/.1) = 26.6 \text{ degrees}$$

Since V = iZ, then the current, i, is found by

$$i = (10/.112) \sin(\Theta - 26.6) = 83.3\sin(\Theta - 26.6)$$
(Current lags voltage.)

The other voltages, then, are as follows:

$$VR = (.1)(83.3)\sin(\Theta - 26.6)$$
$$= 8.33\sin(\Theta - 26.6)$$
$$VL = .250(83.3)\sin(\Theta - 26.6 + 90)$$
$$= 20.83\sin(\Theta + 63.4)$$
$$VC = .200(83.3)\sin(\Theta - 26.6 - 90)$$
$$= 16.66\sin(\Theta - 116.6)$$

Remember, if the circuit is slightly inductive, then the current lags the voltage by a little bit (in this case by 26.6 degrees). VR is always in phase with the current, so it lags the overall voltage by the same 26.6 degrees. VL and VC will lead and lag the current by 90 degrees, respectively. These voltages are graphed in Figure 7-15.

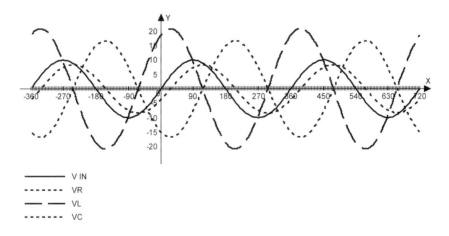

Figure 7-15 Voltages in the RLC circuit.

Looking at Figure 7-15, the overall voltage, Vin = 10sin(Θ), is the curve that goes through the origin. VR, in phase with the current, crosses the x-axis 26.6 degrees later (current is lagging voltage). The curves for VL and VC are each shifted 90 degrees from VR and 180 degrees from each other. That is the first observation.

A second, really interesting observation is this. Consider the voltages at a point, say Θ = 0 degrees. At this point:

$$\begin{aligned}
\text{Vin} &= 10\sin(0) &&= 0 \\
\text{VR} &= 8.33\sin(-26.6) &&= -3.7 \\
\text{VL} &= 20.83\sin(+63.4) &&= 18.6 \\
\text{VC} &= 16.66\sin(-116.6) &&= -14.9
\end{aligned}$$

The latter three voltages sum to the input voltage (zero), as they must.[3] But note that the absolute value of both VL and VC are larger than Vin *ever* is. It is not intuitively obvious that this can happen, but it can. This case illustrates clearly that it can. The total voltage never exceeds the input voltage, because VL and VC tend to cancel each other at every point in time. In fact, the total voltage must always equal the input voltage. But there can be some interesting dynamics *within* the loop when both capacitive and inductive influences are present. And, in particular, the voltage across the inductive and capacitive devices can be considerably larger than the input voltage.

SERIES RLC AT RESONANCE

It gets even more interesting at the point of resonance. Remember that at the point of resonance XL and XC are equal and opposite. So let's look at this example at the resonance point:

$$\omega = \frac{1}{\sqrt{LC}} = \frac{1}{\sqrt{5(10^{-9})(10^{-7})}} = 44.721 \times 10^6$$

---

[3] If you try to replicate these calculations or use a different set of values there are almost certain to be minor round-off errors. Impedance calculations typically must be carried out to a large number of decimal places to eliminate these round-off errors.

$$f = \frac{1}{2\pi\sqrt{LC}} = \frac{1}{2\pi\sqrt{5(10^{-9})(10^{-7})}} = 7.117 \text{ MHz}$$

R = .1 Ω
XL = 5(10⁻⁹)(44.721)10⁶ = .2236
XC = −1/(10⁻⁷)(44.721)10⁶ = −.2236
Xtotal = XL + XC = .2236 − .2236 = 0
Z = $\sqrt{.1^2 + 0^2}$ = .1 (same as R)
Θ = tan⁻¹(0/.1) = 0 degrees

So current, and then the individual voltages are:

i = (10/.1)sin(Θ) = 100sin(Θ)
VR = .1(100)sin(Θ) = 10sin(Θ)
VL = .2236(100)sin(Θ + 90) = 22.36sin(Θ + 90)
VC = .2236(100)sin(Θ − 90) = 22.36sin(Θ − 90)
Vtotal = VR + VL + VC = 10 x sin(Θ)

These voltages are graphed in Figure 7-16.

VL and VC are equal and opposite (180-degrees difference, exactly opposite in phase) at every point in time, so their voltages sum to zero. There is *never* any voltage across the pair of them. But each one individually has a voltage waveform across it that has over twice the amplitude of the input voltage. VR is identical to the input voltage, as it must be, since the other voltages cancel.

You can work out for yourself what happens if R decreases. If R decreases, then current increases. Since there is no reactive component (i.e., X = 0), the current is totally determined by the value of R. So if R were, say, .05 Ω, then the peak current would be 200 amps (200sin(Θ)), and the peak voltage across each reactive component would be about 44 volts. Some strange and exotic things can happen at resonance.

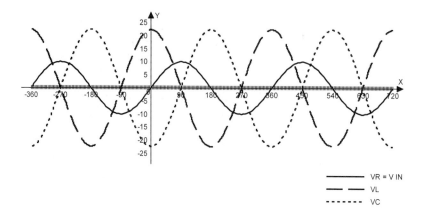

Figure 7-16 Voltages for the RLC circuit at resonance.

ADMITTANCE

In the interest of thoroughness, I add here the concept of admittance. If you are not an active circuit design engineer you probably will never use this term (and I fear even some engineers don't fully understand it). Admittance is simply the inverse of impedance (just as conductance is the inverse of resistance and susceptance is the inverse of reactance; see Chapters 5 and 6). Admittance is designated by the symbol Y and is defined as:

$$Y = 1/Z$$

Ohm's Law applies to admittance (indeed as it must). If $Z = V/i$, then it follows that $Y = i/V$.

Just as impedance is related to resistance and reactance by Equation 7-2,

$$Z = \sqrt{R^2 + X^2}$$

admittance is related to conductance and susceptance by the relationship shown in Equation 7-6:

$$Y = \sqrt{G^2 + B^2} \qquad [7\text{-}6]$$

The phase shift associated with admittance must be the same as the phase shift associated with impedance. The phase shift is associated with the components and their values, not by how we manipulate the definitions of the terms. So, instead of defining the phase shift in terms of resistance and reactance (from Equation 7-4):

$$\Theta = \tan^{-1}(X/R)$$

we can define it in terms of conductance and susceptance (Equation 7-7):

$$\Theta = \tan^{-1}(B/G) \qquad [7\text{-}7]$$

I do not use the term admittance again in this book!

Chapter Endnotes

DETAILED CALCULATIONS FOR FIGURE 7-8

Table 7-2 These Calculations Demonstrate that VR and VC for the Example in Figure 7-8 do Add up to Vtotal.

Angle = A	Vin = 100(sin(A))	VR	VC	Vtotal	Difference
0	0.00	40.00	-40.00	0.00	0.00
26.6	44.78	71.59	-26.81	44.78	0.00
45	70.71	84.85	-14.14	70.71	0.00
90	100.00	80.00	20.00	100.00	0.00
5	8.72	46.82	-38.10	8.72	0.00
10	17.36	53.28	-35.92	17.36	0.00
15	25.88	59.34	-33.46	25.88	0.00
20	34.20	64.95	-30.75	34.20	0.00
25	42.26	70.06	-27.80	42.26	0.00
30	50.00	74.64	-24.64	50.00	0.00
35	57.36	78.65	-21.29	57.36	0.00
40	64.28	82.06	-17.79	64.28	0.00
45	70.71	84.85	-14.14	70.71	0.00
50	76.60	87.00	-10.39	76.60	0.00
55	81.92	88.48	-6.56	81.92	0.00
60	86.60	89.28	-2.68	86.60	0.00
65	90.63	89.41	1.22	90.63	0.00
70	93.97	88.86	5.11	93.97	0.00
75	96.59	87.63	8.97	96.59	0.00
80	98.48	85.73	12.75	98.48	0.00
85	99.62	83.18	16.44	99.62	0.00
90	100.00	80.00	20.00	100.00	0.00

$VR = (100)(.894)\sin(A + 26.6)$
$VC = (50)(.894)\sin(A - 63.4)$

RLC Simulation

UltraCAD has developed a simulation program for use in exploring these relationships further. It is available for download from UltraCAD's Web site. The program will show the waveforms for VR, VL, VC, Vin and Iin for a serial RLC circuit (see Figure 7-17). The program is dynamic in that you can change any one component and watch the waveforms shift in real time. This will give you a perspective on how voltages in a circuit shift in phase as the values of the circuit components change. The instructions for obtaining and running the program can be found in Appendix D.

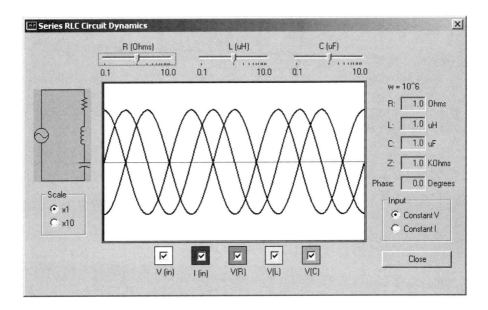

Figure 7-17 UltraCAD's RLC simulation tool.

Imaginary Numbers

In the calculus derivation of the expression $Z = R + jX$, it turns out that the value of j is $\sqrt{-1}$. You should recognize that the square root of -1 does not exist. That is, there is no number that, when multiplied by itself, equals -1.

Electronics is not the only discipline in which this happens. Several other areas of study also have similar expressions, and the common symbol used for $\sqrt{-1}$ is i. In electronics we use j instead so as not to confuse this symbol with the symbol for current.

Expressions of the type $R + jX$ constitute an area of mathematics called complex algebra. Since the square root of -1 doesn't exist, we often call the jX term *imaginary* and we sometimes call complex algebra *imaginary arithmetic*. The two terms making up the expression are called the real part (R) and the imaginary part (X).

See Appendix E for a more in-depth discussion of complex algebra.

We can now look again at why capacitive reactance is given a negative sign. Suppose we have a single capacitor, so $X = XC$. XC is given by:

$$XC = 1/j\omega C$$

If we multiply the numerator and denominator by j:

$$XC = j/j^2\omega C$$

Since:

$$j^2 = \left(\sqrt{-1}\right)^2 = -1$$

this becomes

$$X_C = -j/\omega C = j(-1/\omega C)$$

This fits the format of Equation 7-1. This is where the minus sign for XC comes from.

PART 2

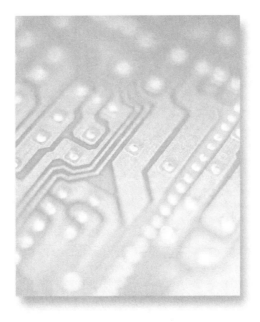

SIGNAL INTEGRITY ISSUES

Ten or 15 years ago most people treated printed circuit boards as totally passive devices for connecting components together. As long as the traces were connected to the correct pins, boards almost never had a negative impact on circuit performance. In more recent years, this is becoming less and less true. Many board designers now need to worry about the parasitic elements of traces (resistance, capacitance, and inductance), the interaction between individual traces, and even the interaction between traces and the outside world. Part 2 explains why this is so and what the designer needs to do about it. Many different topics are covered, including EMI, crosstalk, transmission lines and reflections, and power system decoupling. You will learn what a fundamental role signal rise time has for all of these issues, and then what design rules you need to consider as a result.

8 SIGNAL INTEGRITY OVERVIEW

*O*ne time I served on a panel where I was asked, "At what point does signal integrity become a problem?" The intent of the question was to ask at what *frequency* signal integrity becomes an issue for board designers, but as stated, the question was more general. That got me thinking about how broad the term *signal integrity* really is, even though we have come to think of it in terms of more narrowly defined high-speed issues.

A not-really-facetious answer to the question is that we have a signal integrity problem whenever the signal begins to lose its integrity. And this is not necessarily related to its frequency. Two of the more obvious and common ways a signal can lose integrity is when it becomes distorted or when the signal-to-noise ratio (S/N) begins to degrade.

Signal distortion typically means that the waveform of interest begins to change shape. The degree to which this can happen before it becomes a problem depends very much on the application. Digital signals, which we typically think of as being rectangularly shaped pulses, usually carry one bit of information per clock cycle. They can often withstand a fair amount of distortion without obscuring the bit state they are in. Analog signals, on the other hand, such as we find in video and audio systems, can be very sensitive to distortion. A change in the waveform will often be seen or heard. Your home stereo system, for example, probably has a specification for harmonic distortion, which relates to the "purity" of the audio signal as it is processed.

In my lifetime, home entertainment harmonic distortion specifications have improved greatly, and today's audio and home entertainment systems are, for all practical purposes, distortion free. The most common way to make a distortion-free system is to design it so that the gain is absolutely linear over the frequency range of interest.

S/N issues come into play whenever an unwanted noise becomes detectable or interferes with the signal we are concerned with. For example, if you hear a 60-Hz hum (which is, of course, very low frequency) from your home audio system, you have a S/N or a signal integrity problem. Signal integrity issues and solutions are not new. People who have spent part of their careers designing power or electromagnetic switching systems have understood the importance of decoupling and separation of power supply and grounding systems for decades. Problems associated with reflections, and their transmission line solutions, have also been around for years. Radio frequency engineers (even with "low" AM transmitting frequencies below one MHz) have needed to understand transmission line techniques since the broadcasting industry began. Signal integrity issues at the circuit design level have been around for a long, long time.

In our industry (circuit board design) we have come to equate *signal integrity* with *high speed* only in recent years. Up until now the circuit board has been a purely passive device with virtually no circuit impact (unless you were among the very few designers putting RF or microwave circuits on the board). However, in recent years, the board itself has begun to cause S/N degradation. The reason is almost exclusively because the rise times of the signals we work with have decreased. These decreased rise times, in conjunction with trace inductance and capacitance, now cause noise signals of greater magnitude than in the past.

Let's use inductance for our illustration. Traces have *always* had inductance. That is important for new board designers to understand. Inductance isn't something that just raised its ugly head in the last few years. The voltage generated across an inductor can be approximated by $V = L \times di/Tr$, where L is, in this case, the inductance of the trace (measured in henrys), di is the change in current (in amps) being switched, and Tr is the rise (fall) time (in seconds) associated with the switching current. V gets larger as (among other things) Tr gets smaller. V, in this case, can be considered a noise voltage. Thus, the S/N ratio gets worse as V gets bigger (or, as Tr gets smaller). Therefore, signal integrity gets worse (because the S/N ratio degrades) as Tr gets smaller (i.e., rise time gets faster).

This illustration used inductance. We can use stray trace capacitance to illustrate exactly the same point. There is stray (parasitic) inductance and capacitance all over our boards.

So, back to the initial question: At what point does signal integrity become a problem? For board designers, the answer is: When the rise time decreases to the point where the parasitic inductances and capacitances on the board begin to result in noise signals and transients that become troublesome. When is that? Well, it depends, of course, on the circuit specifics, so it is very difficult to generalize on a specific value. Some typical symptoms that might accompany signal integrity problems might include:

1. Design fails FCC compliance testing (a sure sign!).
2. Design fails intermittently, often without discernable pattern.
3. Design *did* work reliably, now it doesn't, after:
    Nothing.
    Changing IC supplier.
    Changing board fabricator.
4. Boards must be selected to work together.
5. Design is sensitive to power supply variations.
6. Design is sensitive to temperature variations.
7. Design works in the lab, but is intermittent in the field.

Symptom 3 is a particularly subtle problem, but one that is becoming increasingly common. Consider the following situation: A component (IC) supplier improves its processes. The process results in smaller die, better yields, lower cost, and improved performance, especially with regard to rise time. No announcement is made regarding these improvements, which are introduced in-line without fanfare or part number modification. It is hard to be critical of this—after all, everybody is a winner here, right? Well, almost everyone.

Consider a design that has been proven over the years and one that has been highly reliable. Suddenly, reliability problems begin to appear and they are difficult to diagnose, in part because they are so unexpected. Sometimes problems like this have been traced to unannounced improvements in the designs of the components used. The faster rise times caused signal integrity problems that were not there before. Board design techniques that were acceptable with lower rise time circuits were no longer acceptable. Similarly, problems like this can occur when switching from one component supplier to another. The parts may be considered substitutes from the standpoint of form, fit, and function, but may have subtle rise-time or timing differences that cause them to react differently in circuits.

Circuit board materials can also play a significant role in signal integrity. Factors like board thickness are obvious. But many people don't understand the important role that other parameters, such as relative dielectric coefficient, can play in controlling signal integrity. If a company changes board fabricators, and appropriate control over fabrication parameters, including relative dielectric, is not maintained, unexpected problems can be introduced during this type of changeover.

In general, signal integrity problems on circuit boards fall into four areas, all of which are related to rise time (or related frequency harmonics). They are:

- EMI (radiations beyond the board or susceptibility to radiations from outside the board)
- Reflections on a single net
- Crosstalk between two or more nets, in many ways a special case of EMI
- Power system stability during component switching

We are going to talk about each of these areas in this part of the book. We will cover the nature of the problem (i.e., what causes it), how the problem is manifested in circuits, and what design solutions can be employed to minimize its effects.

You may be surprised at the number of solutions we are going to offer. There aren't very many, because, in part, there aren't very many *causes* to the problems. The primary cause is the decreasing rise (or fall) time of our signals and its interaction with the reactive characteristics of the stray inductances and capacitances on our boards.

Thus, you will notice that the same figure sometimes pops up in several different chapters. There are certain design "errors" we can make that have multiple consequences. For example, a slot in a plane can cause EMI problems (Chapter 9), reflection issues (Chapters 10 and 11), and crosstalk problems (Chapters 12 and 13).

One aspect to the solution to signal integrity problems is usually a real eye opener for designers just learning about these issues. Three of the important electronic principles covered in Part 1 of this book are:

1. Current is the flow of electrons.
2. Current flows in a closed loop.
3. Current is a constant everywhere in a closed loop.

This has fairly basic consequences for designers:

1. Every signal has a return.
2. You need to know where it is!

When you stop to think about it, designers often spend an enormous amount of time and effort planning and strategizing how to route 50% of the signals, and then leave the other 50% (the returns) to chance. *One of the primary causes of signal integrity issues is the lack of control over signal return currents.* This will be a common theme in almost every chapter.

Control over signal return currents is almost totally achieved by using planes. One solution that's common to every area is the use of reference planes. It is pretty much axiomatic that (for most of us) good signal integrity performance *requires* reference planes. Planes play an integral role in:

- Controlling EMI
- Stabilizing trace impedance, an integral part of controlling signal reflections
- Controlling crosstalk
- Providing high-frequency decoupling for our power systems.

Another solution that is common to at least two areas is the minimization of the distance between the trace and the reference plane (i.e., the height of the trace above the plane). Minimizing this distance (a) minimizes EMI and (b) minimizes crosstalk.

A third solution relates to the general problem of minimizing inductance. This is especially important around connections to devices (pads, vias, soldered connections, etc.). Minimizing inductance is generally even more important around power and ground planes.

A few other solutions will be offered at various specific places. But the point remains. It may seem like there are a lot of potential signal integrity issues with our boards, but there are only a few causes. Therefore, there are only a relatively few required solutions. You will find this out by the end of the book.

A student in one of my classes once observed (and I am paraphrasing here), "I can look at boards from (and he named some large computer and memory manufacturers) that seem to violate everything you and the other instructors are teaching in these seminars. How come?" This is an excellent question. To start with, all the potential signal integrity problems highlighted in this book and in the various seminars are just that, *potential* problems. The rules we provide are ways to solve these potential problems if they do occur. Will the problems always occur? No, but they often do. Will these solutions always work? Usually, but not every single time.

Here is another viewpoint: If (and this is a big if) you are thoroughly familiar with every aspect of every signal on the board, then you will know, for example, if two traces will or will not crosstalk to each other during operation. Therefore, you will know whether the crosstalk guidelines need apply to these particular traces. If you have control over every aspect of the design of the system, then you can optimize the pin outs from the various devices so that signal integrity problems are minimized. If you have unlimited resources, you can have several people make several design iterations, testing different concepts, until you find an optimum set of rules for your particular design (recognizing that several of the iterations are likely not to work well at all).

A friend told me once that his company went through *thirteen* design iterations before they finalized their board design for production. He proudly explained that the last 12 were just "tweaks." After all, the system *did* boot up with the first try. They had the resources to keep trying alternative strategies until they truly optimized the design.

But if you live in the more real world that the rest of us live in, then for all practical purposes you have only one or two shots to get the design right. In that case it is often easier to simply follow good design rules, just to be sure. A less-than-optimum design is usually more cost effective than multiple design iterations. If compromises must be made, knowing the proper design rules will help you troubleshoot a board later to determine which compromises turned out not to work and need to be corrected.

We often make trade-offs in our designs. That is not uncommon, and it is acceptable. What's important is that we do so with our eyes open, so we are alert to potential problems in the future and how to correct them. It is for these reasons that studying books like this and taking as many seminars as you can are important.

9 ELECTROMAGNETIC INTERFERENCE (EMI)

Background

*C*urrent is the flow of electrons. When electrons move down a trace or a wire, current flows. Electrons are negatively charged particles. Therefore, we can envision that there is an electrical (negative) charge around a wire as current flows, and the magnitude of this charge depends on the number of electrons (magnitude of the current). We call this charge an *electric field*. Furthermore, as the current flow changes in magnitude, this electric field changes in intensity.

Also, as electrons flow, a magnetic field is generated around the wire or trace. Andre Marie Ampere, for whom the measure of current is named, is credited as the first person to state this law. This is the basic relationship behind an electromagnet. The strength of the magnetic field is related to the magnitude of the current flow, so the magnitude of the magnetic field will change with changing current levels.

Therefore, around any wire or trace carrying a current, there will be an electrical field (often designated by the letter E) and a magnetic field (often designated by the letter H). Together, these two fields form an electro<u>m</u>agnetic field, as described in Chapter 2. The two components of the field (electrical and magnetic) must flow together at the same speed. The speed at which they can travel is determined by the medium they travel through, the medium surrounding the current flow, and specifically the relative dielectric constant of that medium. Thus, the propagation speed of a signal is determined by how fast the electromagnetic field can travel through the surrounding medium (see Chapter 2).

In 1831 Michael Faraday published Faraday's Law of Magnetic Induction. This law states that a *changing* magnetic field (itself caused by a changing current flow) can induce an electrical current in an adjacent wire. So if we send a changing (AC) current down a wire or trace, it can induce a similar current in an adjacent wire or trace. This can be a good thing: It is the basic principle behind radio and television transmission and reception. But it can also be a bad thing: It is the cause of crosstalk and FCC compliance testing problems. As board designers, our challenge is to design boards that maximize this effect when we want to transmit a signal and minimize this effect when we don't. Transmissions we *don't* want are called EMI (in this chapter), or crosstalk (in Chapters 2, 12, and 13.)

In one sense, all of our traces are antennas. A good transmitting antenna is also a good receiving antenna. Therefore, any trace that is a good "radiator" is also a good receiver. Designs that *emit* EMI are also more *susceptible* to EMI. In this chapter we will look at design techniques that help reduce EMI. We will see later that the very same techniques also help minimize crosstalk.

FIELDS AND CANCELLATIONS

Figure 9-1 illustrates the magnetic field lines around a wire when the current is flowing out of the page. The direction of the magnetic flux lines can be determined by the right-hand rule. Point the thumb of your right hand in the direction of the current flow. The magnetic flux lines curl in the direction of your fingers.

Figure 9-1 Magnetic flux lines curl around a wire following the right-hand rule.

Now consider Figure 9-2. Here there are two conductors, one with current coming out of the page and one with current going into the page. If these two wires are carrying a signal and its return current, then the currents will be equal and opposite. Therefore the magnetic fields will also be equal and opposite. Think of yourself as positioned off to the right-hand side of this page. One of these wires will be closer to you than the other (the one with current going into the page). Thus the EMI radiation from that wire will be greater than that from the other. You will be able to measure this EMI radiation, and under the right conditions this radiation might be significant.

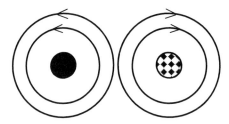

Figure 9-2 A signal and its return have magnetic fields that approximately cancel.

But if we move the two wires closer together, then you, as an observer off to the right, would begin to see radiations from each wire that are more equal. And since they are opposite in phase, or polarity, they tend to cancel. In the limit, if the wires were very close together, their EMI radiations would completely cancel and would not be measurable. Herein is one of our most basic principles: *If you want to minimize EMI radiation, keep the signals and their returns close together.* That, of course, may be easier to say than to do.

This is why wires formed into a twisted pair work fairly well in AC environments. Consider Figure 9-3. The twisted pair of wires carries a signal and its return to and from an operational amplifier. Again, position yourself off to the right of the page. At any given point along the twisted pair, one wire will be closer to you than the other, so the EMI radiation from it will be slightly larger than from the other. But since the wires are twisted, on average they are the same distance from you, so on average, the radiation from each one cancels the other. Therefore, you receive very little net radiation.

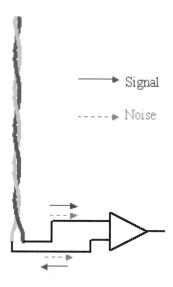

Figure 9-3 A twisted pair has desirable features from an EMI standpoint.

Of course, the wires in a twisted pair can act as an antenna and receive noise, also. Twisted pairs usually feed into differential circuits that are (a) very sensitive to the difference in signal between the pair, but (b) relatively insensitive to any (common mode) signal that exists on both wires. The common mode rejection specification is one measure of a circuit's insensitivity to common mode signals. We explore this more fully in Chapter 14.

SOME BASIC "TRUTHS"

There are a few truths you should always keep in mind. The first, and perhaps most important, is that current always flows in a closed loop. Remember, current is the flow of electrons, and if current does not flow in a closed loop, where do the extra electrons come from or go? Because we know we don't have to deal with buildups of stray electrons on our boards, it seems reasonably intuitive that currents really do flow in a closed loop.

That means every signal has a return, somewhere. And you (as the designer) need to know where that is. It sometimes amazes us to discover that we designers spend an inordinate amount of time worrying about, considering, and strategizing about 50% of the signals, and then leave the other 50% (the return signals) somehow to chance. So the first truth is that every signal has a return and you need to worry about and control where it is.

A second rule is that the return current will always follow the path of least impedance. We all have heard the rule that signals follow the path of least resistance, but that is a special case, not the general one. It only applies to low-frequency signals. The general rule, for all signals, and particularly the high-speed signals we worry about, is that return currents follow the path of least impedance. This may not always be where you think it is.

SIGNAL COUPLING

Consider Figure 9-4. A changing current i1 is flowing down Trace 1. The changing current creates a changing magnetic field around Trace 1. That changing magnetic field couples into Trace 2 with a coupling coefficient whose value is k. The coupling generates a current in Trace 2 whose value is k x i1 and whose direction (by Lenz's Law, see Appendix B) is opposite to i1. This is exactly how a transformer works, and is the fundamental basis behind all transformer designs.

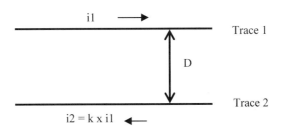

Figure 9-4 Traces can and do couple into each other.

The coupling coefficient will have a value somewhere between 0 and 1.0. Zero means no coupling whatsoever, which can occur if (a) the traces are very far apart, or (b) the current is not changing (i.e., is DC). A coupling coefficient of 1.0 means perfect coupling, which of course cannot be achieved. The value of k, for our purposes, depends primarily on (a) how fast the current is changing, or the rise time (di/dt), and (b) how closely the traces are spaced (D in the figure). Fast rise times and closely spaced traces lead to higher coupling coefficients. Realistic values for k on circuit boards where the traces are close together are in the range of .4 to .6.

Consider the special case where i1 is a signal and i2 is the signal return. Therefore, i2 = –i1. The coupled signal, k x i1 actually helps boost the return signal. But the return signal, i2, also couples into Trace 1, boosting the primary signal, i1. So, the mutual coupling of the signal and its return work to reinforce, or boost, each other. We can interpret the effect of this two ways: (a) It takes less force (voltage) to push the same amount of current through the circuit, since these currents reinforce each other; or (b) more current can flow for the same voltage, since reinforcing currents are generated on each trace. By Ohm's Law, V = iZ. No matter which interpretation you prefer, it means that the impedance goes down when there is coupling of this type. Furthermore, the impedance gets progressively lower as the coupling increases, or as k increases. And k increases as (a) the distance between the traces decreases, or as (b) the rise time decreases (gets faster).

Now consider Figure 9-5, which illustrates the simple case of a microstrip trace placed over a plane. Assume there is a current flowing down the trace and the return current is on the plane. A good question is: "Where is the return current?" Assume the return current is at Point a. Following the preceding discussion, there will be some mutual coupling between the signal and its return current, resulting in some mutual reinforcement. Now let's assume the return current is, instead, at Point b. Here, the coupling is stronger, because the signal and its return are closer together. Since the coupling is stronger, the overall impedance will be less. Therefore, Point b is a lower impedance path than Point a. Point c represents a signal return path that is even closer to the signal, so the coupling will be even stronger at Point c and the impedance will be even less. Point c represents the closest point between the signal and its return—it is directly under the signal trace—and therefore the point of maximum coupling and lowest impedance.

This is why high-speed signals want to return directly underneath the signal trace: It is the path of lowest overall impedance. Be careful to note, however, that this argument depends completely on the coupling coefficient between the two signals. The coupling coefficient depends, among other things, on di/dt, or rise time. If there is no rise time (e.g., the signals are DC power signals) there is no coupling and the return signal might be anywhere. In fact, the DC return will truly follow the path of least resistance anywhere on the circuit board.

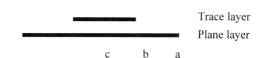

Figure 9-5 Cross-sectional view of a trace and underlying plane. A return current will "want" to follow the path of least impedance, Path (c).

To summarize, DC signals might return anywhere on the board. But AC signals want to return directly underneath the trace carrying the signals. This tendency increases as the rise time decreases (gets faster).

LOOP AREA

Consider a simple configuration with a trace directly over an uninterrupted plane. The signal travels down the trace and the return signal comes back on the plane directly under the trace. Conceptually, we can define a loop that the signal travels as being the length of the trace from the driver to the receiver, down through the receiver to the plane, back under the trace, and back through the driver again. In simple terms, the area enclosed by this loop would be the length of the trace multiplied by the height of the trace above the plane (Figure 9-6). We call this the *loop area* of the signal. The loop area is the area defined by the signal as it travels down the trace and returns back to the source. (This is one reason you need to know where the return signal is on your boards.)

Now loop area is important for this reason. From a practical standpoint, for high-speed signals, EMI is related to loop area. *If you want to minimize EMI, you must minimize loop area.* It's as simple as that. Most of us are intuitively aware that traces, even microstrip traces, routed close to planes generally perform well from an EMI standpoint. That's because their loop areas are small. For the next few pages we'll look at illustrations where loop areas might get out of control, perhaps unexpectedly so.

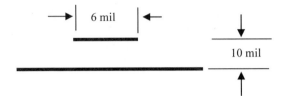

Figure 9-6 Loop area, in this illustration, is 10 x length mil^2.

Slots in Planes 1 Figure 9-7 is an illustration you will see three times in this book. There are many reasons you don't do what is shown there. The figure shows a trace crossing a slot or discontinuity in the plane.

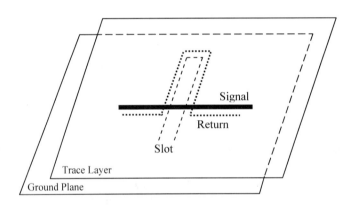

Figure 9-7 If a trace is routed across a slot, the return current must go around the slot, creating a larger loop area.

What might cause a slot or discontinuity in a plane? Perhaps you have finished the design and then your engineer comes to you with "one more little part" that needs to be added. Making provisions for the routing of this new part may involve a significant amount of rerouting of already routed nets. But, perhaps you could just put a little slot in one of the planes, route one or more of the new traces in that slot, and finish the board quickly. Or, perhaps the plane has been partitioned in some way, creating a discontinuity. Or, perhaps there is a special component requirement nearby that results in a small discontinuity.

No matter what the cause of this slot or discontinuity, if a high-speed trace is routed over it, the return signal will be unable to travel underneath that trace. When the return signal hits the discontinuity, it must travel around it and then come back to its position under the trace again. This trip around the slot or discontinuity creates a loop. Since EMI is related to loop area, you have now created a possible EMI problem where none existed previously.

Return Pathways Figure 9-8 illustrates a case where two boards are connected through a connector. The signal between IC1 and IC2 travels through connector Pin A. What provision should be made for the return signal?

If we use Pin B for the return signal, the loop area will be pretty small. EMI performance should be satisfactory. But if we use Pin E for the return signal, a significant loop will result. This loop may cause an EMI problem.

This illustration highlights the importance of pin assignments in our devices and connectors. It should also be noted that some connectors are better than others with regard to their internal paths. Some poorly designed connectors introduce loop areas even between adjacent pins.

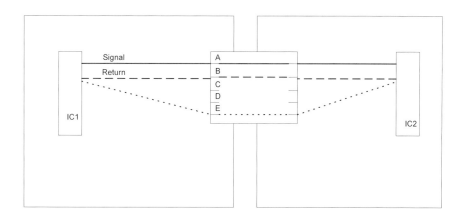

Figure 9-8 If provision is not made for a signal return (ground) through a connector, a larger loop area might be created.

Figure 9-9 illustrates another common problem. The figure shows a row of clearance holes for pins for perhaps a component or connector. In Figure 9-9a and

Figure 9-9b, a signal trace connects to one pin and the return (ground) pin is on the back side. In Figure 9-9a the clearance holes are wide enough to remove all the copper from the plane under the connector or the device. The return signal, traveling under the trace, must then travel around the copper void area to get to the ground pin. This creates a loop that might become an EMI problem.

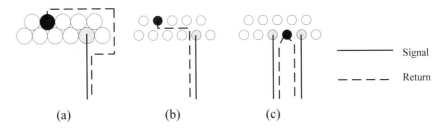

Figure 9-9 Good practice includes providing copper paths between through hole pins and assigning pins to minimize loop areas.

A better solution is shown in Figure 9-9b. The clearance holes are smaller, so that they do not remove all of the copper from the plane in this area. Now there is a path for the return signal to travel between the pins to the return (ground) pin. This approach is marginally better than Figure 9-9a and results in a smaller loop area. And this illustrates why it is usually preferable to provide a copper path on the planes through all areas like this.

However, Figure 9-9c shows an even better solution. If you have the flexibility to do so, it is best to assign pins so that there is always a ground (return) pin beside every signal pin. This provides the absolute minimum loop area for the signal and its return, resulting in the best EMI performance.

Power Plane Returns Now is the time to introduce an interesting complication to all this. We earlier pointed out that the return signal will follow the path of least impedance. That is the point on the plane closest to the signal trace. What if that plane happens to be a power plane rather than a ground plane? How does the return signal get back from the power plane to the ground pin of the device?

First, DC voltage has no meaning to a high-speed return signal. The return signal can be just as happy on a power plane as it is on a ground plane. From an AC standpoint there is no difference between a power and a ground plane. The most obvious evidence of this point is all the bypass capacitors we have connecting the two planes together. From an AC standpoint the planes are shorted together all over the place.

So, what happens if the return is on a power plane? The exact answer may not be satisfying or may not even be known, but the practical answer is that the return signal probably transitions from the power plane to the ground plane through a nearby bypass cap. And if we have very high frequency harmonics, the transition may be through the planar capacitance between the two planes themselves, if such planar capacitance has been designed into the board.

It may not be a satisfying answer that the signal passes through a nearby bypass cap. After all, that leaves the definition of the loop area somewhat up to chance. I have heard a few people, for example Howard Johnson and Henry Ott (two highly respected experts in the field), sometimes advocate the placement of bypass caps near connectors or devices such as that shown in Figure 9-9 simply to provide a path for the return current. The bypass capacitor would not be intended to provide a decoupling function at all. Its sole purpose would be to control the loop area for the return currents on the power plane and therefore control EMI.

A few people advocate never routing such a signal adjacent to a power plane. They advocate routing all such signals adjacent (referencing) to only ground planes. Most people (including us at UltraCAD) consider this position too extreme. If your engineer (customer) insists on this constraint, however (after all, the customer *is* the boss), EMI and loop area are two possible reasons.

Changing Trace Layers Figure 9-10 illustrates another interesting case. Figure 9-10a shows a simple signal referenced to a plane. Figure 9-10b illustrates the trace transitioning to the other side of the same plane. In this case the return signal would simply transition to the other side of the same reference plane, and there should be no effect on loop area.

What about the case shown in Figure 9-10c? Here the signal transitions to an entirely different trace layer and reference plane. A legitimate question is, "How does the return signal get back under the trace on the new signal layer?"

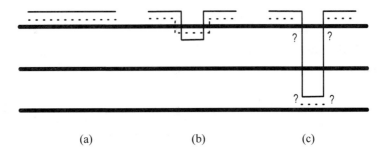

Figure 9-10 Some engineers are concerned about the location of the return path, and possible loop area problems, if signals are allowed to transition to different trace layers.

Again, the not very satisfying answer is through a nearby bypass cap. Again, some people advocate placing a bypass cap near each via for the single purpose of providing a path for the return signal and reducing loop area. A few circuit design engineers even prohibit transitions of traces from one trace layer to a completely different one because of the potential for increased loop areas and EMI problems (although most people would call this an extreme position to take).

On the designs we do at UltraCAD there are typically quite a few bypass capacitors. We generally take no particular special precautions when transitioning signals to different trace levels, nor do we pay particular attention to which plane (power or ground) the return signal is on. We are not aware of any customer design failures that can be traced to either of these practices.

Unrelated Planes Many boards have separate, well-defined areas on them for different power systems. There may be both 3-volt and 5-volt power supplies, for example, or perhaps separate analog and digital power supplies (or even transmitter and receiver power supplies, etc.). Often, in such cases, we define planar areas for these supplies.

Figure 9-11 illustrates a situation in which there are separate analog and digital power supplies and planes. Consider a trace that must be routed between the digital ICs 1 and 2. Ideally, the trace would be routed exclusively over the digital power plane.

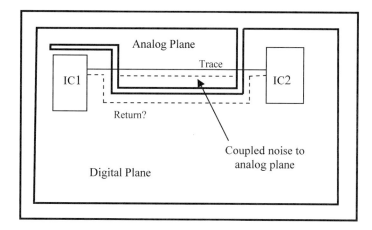

Figure 9-11 Never route a trace over an unrelated plane.

What if we took the shorter path and routed it over the analog plane? The question then is, "Where is the return signal?" There are two possible answers to that question, both of which are bad.

One possible answer is that the return signal stays on the digital plane and flows around the analog plane. But this creates a wider loop area and therefore potential EMI problems. Another possibility is that the return signal somehow finds a path onto the analog plane and continues on under the trace. We may not have an EMI problem in this case, but what we do have now is a digital signal on the analog plane. This digital signal might interfere with other analog signals in the area and create a crosstalk or ground noise problem. After all, the whole reason we have separate power systems is to keep the analog and digital signals separated in the first place. Either of these possibilities is bad. So the right answer is: *Don't route traces over unrelated planes.*

Stripline We intuitively know that a shielded wire (e.g., a coaxial cable) performs well from an EMI standpoint. We probably are aware that another defense against EMI is a shielded enclosure. So we can understand that traces shielded between planes probably perform well, too. This is true, within limits.

The single most important rule for EMI control is to control loop area. Even in a stripline environment (where traces are shielded between two planes) a large loop area can cause EMI problems both externally (EMI radiation) and internally (crosstalk). But all other things being equal, a stripline environment adds an extra measure of EMI control. So, even though microstrip traces can be designed to perform fully satisfactorily from an EMI standpoint, designers sometimes opt to place all critical signal traces in a stripline environment for an added measure of safety.

STUBS

Stubs are short trace segments that extend, unterminated, from (an often controlled impedance or transmission line) main trace. Most commonly they connect a pad, and therefore a device, to a net. Figure 9-12 illustrates four stubs, S1 through S4. S1 and S3 are normal stubs that we all use. Stub S2 is too long and should be avoided in high-speed designs. Route the trace closer to the pad as shown for Stub S1.

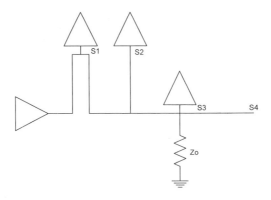

Figure 9-12 Various examples of stubs.

There are two reasons for avoiding long stubs. One relates to transmission lines and is covered in Chapter 10. The other relates to EMI.

Look at your cell phone. There is probably a short antenna extending out from the top. Now look at Stub S2 in Figure 9-12. This unterminated stub looks a lot like the antenna that extends out of your cell phone. Stubs like this can radiate just like antennas and should be avoided.

The stub shown as S4 is particularly harmful. This might have been created when a trace was rerouted and the designer forgot to remove a short segment from the original route. This *is* an antenna. Its results might not be harmful, depending on the circumstances, but they are never beneficial. Never allow such stubs to exist on your board.

Antenna theory is a very complex area that is not well understood by most people. There are a huge number of variables that affect whether a wire or trace can function well as an antenna. Some of these include the environment the wire or trace is in—the degree to which it is shielded by the planes, its length, the surrounding material and its relative dielectric coefficient, the specific shape of the wire or trace, and so on. Another variable is the frequency of the signal (or its harmonics). So it is difficult to look at a stub and say with certainty whether it will create a problem. But the general rule is that stubs create problems, so minimize them. Be sure to take the time to trim back any trace that is left "hanging" and not connected to anything.

COMMON MODE

Loop areas are one of the two most common causes of EMI. The other cause is referred to as common mode problems. The problem is that common mode is a poorly understood and frequently misused term. Before we can talk about this part of the EMI issue, we must spend a little time talking about the term and its misuse.

One of the most frequent explanations of common mode is made in contrast to another term, *differential mode* (not to be confused with differential traces or signals). Differential mode is the mode we are usually familiar with. That is, a signal travels down a well-defined path and returns, in the opposite direction, on another well-defined path. Since the signal and its return flow in opposite directions, the term differential is used. Following this logic, *common mode* then refers to the case where the signal flows in the same direction on both the trace and the return.

Since all signals, by Kirchhoff's Laws, must flow in closed loops, this explanation becomes troublesome when you really start to think about it. If common mode currents flow in the same direction, how do they return to their source? Then there is the confusion caused by the term differential mode in this context, and differential mode in the context of differential signaling and differential traces. We will find later (in Chapter 14) that the term differential mode is actually misused in the latter case.

We are going to leave the differential signal case to Chapter 14 and talk about common mode EMI issues here only in the context of a single-ended, individual trace. But the extension of the problem to the case of differential traces, and then how to deal with it, will be somewhat apparent, even if the actual causes of the problem will not.

Let's define common mode a little differently from the definition we are normally familiar with. Let's use the term differential mode (in the context of a single-ended trace) to mean the currents that flow where we expect them to flow based on the schematic and the layout. If we are worried at all about EMI, we have already learned that we want to minimize loop areas and route traces directly over, and close to, reference planes. Therefore, the differential mode signals travel down the trace and return on the plane directly under the trace.

Common mode currents are those that flow anywhere else.

This raises several obvious questions. (a) How can this happen? (b) Even if it happens, what's the problem? (c) What's the relationship between this definition and the normal one generally used? (d) What, if anything, can be done about it?

How Can This Happen? Figure 9-13 illustrates one example of common mode currents based on our definition. A signal travels down the trace and returns on the ground plane directly underneath the trace. This is the differential mode signal current, i_d. Another part of the signal travels down the trace and returns through a completely different unintended path. This is called the common mode signal current, i_c.

This can happen for several reasons. One is that the ground plane is not a perfect conductor. There is some inductance (admittedly small) associated with it. There can be a voltage divider action involved where the return current splits along the intended path (i_d) and some other paths (i_c).

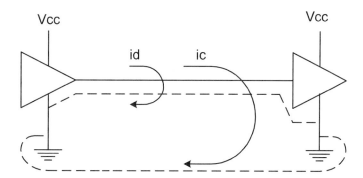

Figure 9-13 Differential mode, id, and common mode, ic, currents.

Or, perhaps there is a wire, stub, or cable shield (e.g., a coaxial cable) attached to the trace. Very fast rise-time transients may travel along such paths and radiate to other traces or surfaces. The characteristic impedance of a wire in space is generally known to be on the order of 370 Ω, which means the impedance along another unintended path could be in this range.

If the impedance along the plane is a fraction of an ohm and a radiated signal is on the order of 370 Ω, the amount of current flowing in the alternative path would only be a tiny fraction of the primary (differential) current. But it would not be zero. And that is the point. There can be unintended (although very small) common mode currents flowing in places we don't expect.

So What's the Problem? We have already learned in this section that to control EMI we want to minimize loop areas. The problem with common mode currents is we don't know where they are flowing. They may be flowing in very (relatively) large loop areas. If so, they may be radiating very significantly. So a common mode signal may only be a tiny fraction of the magnitude of a primary signal, but if it is radiating many times more loudly, it can still become a significant EMI problem.

Relationship to Conventional Definition Figure 9-14 illustrates the conventional definition of common mode current (ic). It is usually shown as appearing on both the trace and ground, flowing toward the receiver.

First, there is no necessary reason that it flows toward the receiver; common mode could just as easily exist on the trace and plane flowing away from the receiver (though we might have trouble envisioning how such a signal might be generated).

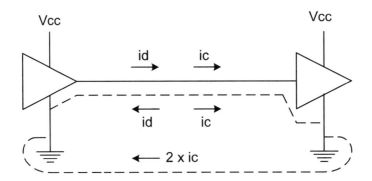

Figure 9-14 Any signal can be separated into differential and common mode components.

We suggested that common mode current is any current that flows in the circuit along a path other than the one we intended. *Any* combination of signal and return current can be broken into differential and common mode components suggested in Figure 9-14, assuming common mode currents exist at all. For example, assume that the signal current on the trace is 10,000 μA (10 mA) and the return signal on the plane under the trace is slightly smaller, say 9,950 μA. This, then, means that there is 50 μA flowing back to the driver along some other path. The differential and common mode components of Figure 9-14, therefore, under these assumptions, are id = 9,975 μA and ic = 25 μA.

What Do We Do About It? Since common mode currents are unintended, it seems like they would be difficult to control. Indeed, that can be the case. There are a few basic ground rules for controlling common mode problems:

1. Use good high-speed design techniques everywhere in your design.
2. In particular, maintain a solid, continuous reference plane for every trace.
3. Common mode problems are smaller for stripline traces contained between planes. If possible, route all critical traces in stripline environments.

Chapter 9

4. Minimize the presence of any stubs or extraneous trace lengths associated with any high-speed trace. Try to eliminate any path for a signal to flow on other than the intended path.
5. Since all signals ultimately flow from the power plane through a driver to a receiver to the ground plane and back to the power plane, minimize this path length. One way to do this is to use power and ground plane pairs in your board stackup. Then ensure there is good capacitive decoupling between these two power sources.

THE 20-H RULE

A rule that is sometimes proposed has been nicknamed the 20-H rule. It has been credited originally to W. Michael King and a thorough explanation can be found in one of Mark Montrose's books.[1] The rule argues as follows: If there are high-speed currents on the board, there are electromagnetic fields associated with them. At the edge of the planes (presumably at the edge of the board) these fields will fringe outward from the board as shown in Figure 9-15a. But if one of the planes is recessed, as the power plane is in Figure 9-15b, the fringing tends to be directed downward to the other plane instead of outward. Thus, outward EMI radiation is reduced and there is less chance for an external EMI problem.

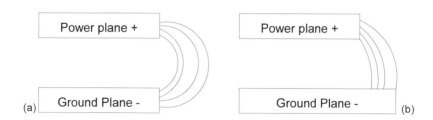

Figure 9-15 Recessing a plane (b) may reduce outward fringing.

The reduction in fringing is directly related to the amount one plane is recessed from the other. The argument goes that the reduction in field intensity is about 70% if the recess is 20 times the distance between the planes (H); hence the name of the rule.

---

[1] *Printed Circuit Board Design Techniques for EMC Compliance*, Montrose, Mark; IEEE Press, 1996, p. 26.

A few studies have tried to examine or measure this effect. In general they all seem to come to the same conclusion: The 20-H rule is not particularly effective.[2]

There is a certain intuitiveness to this conclusion. The high-speed signals are carried on the traces. On well-designed boards, the signal returns are confined to the planes directly under the traces. Therefore, there should not be a great deal of radiation at the edge of the board anyway, assuming we have done everything else well. At UltraCAD we have never used the 20-H rule on a board and we have had no particular EMI problems. Our position is that if this is the worst problem you have with your design, you are already in pretty good shape.

PICKET FENCES (FARADAY SHIELDS)

Some people advocate placing closely spaced vias between two ground planes to form a type of shield around the perimeter of the board. This technique is sometimes called a *picket fence* because of the obvious resemblance. The idea is that if any stray EMI is generated within the board, it can't get past the shield formed by the vias and radiate into the outside world.

People differ on how closely the vias should be placed, but you can hear estimates as close as 1/20 wavelength (which could be roughly every inch for a 1 ns rise time signal). Wavelength is discussed in Chapter 2.

Not all engineers like the idea of picket fences. UltraCAD has never made such a design for any of our customers. A picket fence isn't needed if there is no EMI generated in the first place. My philosophy and advice is to concentrate design efforts on preventing EMI generation in the first place, rather than in controlling it after it is generated.

---

[2] See, for example, "Effects of 20-H Rule and Shielding Vias on Electromagnetic Radiation From Printed Circuit Boards," Chen, Huabo and Fang, Jiayuan, Dept. of Electrical Engineering, University of California at Santa Cruz. They conclude, "For the two-plane structure, 20-H rule yields much more radiation than the normal structure. For the multiple plane case, no significant change in radiation is found if the 20-H rule is applied to the internal planes."

10 REFLECTIONS AND TRANSMISSION LINES

COMMUNICATIONS MODEL

*F*igure 10-1 illustrates what we sometimes call a general communications model. Any time there is a communication, we can think in terms of there being (a) a sender of that communication, (b) a message that is being sent, (c) the media over which the message is being sent, and (d) the (intended) receiver of the message.

Figure 10-1 A general communications model.

For example, at this moment you are reading something "sent" by Doug Brooks, author of this book. The message is what is written on this page. The media is the page you are reading, and the receiver is you, the reader.

On PCBs, the sender is the driver, the message is typically the change in state from a high to a low signal or from a low to a high signal, the media is the trace, and the receiver is the receiving circuit.

The evening news is "sent" by the news anchor (and his or her team behind the scenes). The message is what you see on TV, and might be designed to be entertaining as well as informative. The media can be thought of in simple terms as the television system. In reality it can be a much more complex system of microphones, cameras, switchers, satellites, retransmitters, and so on. The receivers are those watching the program.

In advertising, we think of the sender being the advertiser (e.g., Nike). The message is the advertising message. The media is the media chosen. The receivers are those targeted to receive the message. When we think about advertising in this context, we begin to think about some of the issues that advertisers face:

- How do you select which receivers to target?
- Where are they?
- What media will reach those targeted receivers most efficiently?
- How do we structure our message so that the receivers "hear" it while being bombarded with thousands of other messages at the same time from other senders trying to get their attention?
- How do we make sure the receiver "hears" the same message the sender thinks he or she is sending? It is fairly common for targets of advertising messages to perceive a different message than that which the advertiser intends to send.

Think about this communication problem: Picture yourself sitting down at the end of a high school gym at assembly time. The football coach is at the other end handing out the team awards. He is speaking through a handheld public address (PA) system. The floor is made of hardwood, the walls are concrete, the ceiling is metal, and the echoes are awful. It is almost impossible to understand what the coach is saying. The media is the primary cause of this particular communications problem.

Now, this time, picture yourself positioned along a PCB trace. The driver is sending a signal down the trace and you are trying to "hear" it. But here is the problem: Any time a signal travels down a wire or a trace, it reflects. Every time—(well, almost). Notice what I said and what I didn't say. I did not say "often" reflects. I did not say "sometimes" reflects. I did not say "under certain circumstances" it reflects. It *always* reflects. (Well, it turns out there is one very special case when it doesn't.

Chapter 10 177

But this special case doesn't happen by accident, as you will see. It is this special case that is, in fact, the point of this chapter.)

So, in both our examples, the high school gym and the PCB trace, there are reflections and echoes that can interfere with the receiver understanding the message being sent. Here are some solutions to these communications problems:

1. We can encode the message, so we can pull it from the noise level. There is an entire science related to encoding and decoding messages in noisy environments, and this certainly a legitimate solution. It's not particularly practical in the gym environment, however, and probably not in our PCB environment, either.
2. We can listen "harder," or become better listeners. In the gym, we can concentrate harder on what the coach is saying. On our boards we can use better, more selective, but probably more expensive receivers. Again, these are legitimate solutions, but perhaps not desirable.
3. We can shorten the distance the message travels. In the gym, we can get up and move closer to the coach. We can shorten the trace on the board. On the circuit board, this is usually a highly desirable alternative, but we probably have already designed the traces about as short as they can be, so further shortening them is usually no longer an alternative.
4. A particularly interesting solution is to slow down the message. If the coach will speak more slowly it will clearly be easier to understand him.[1] The analogy on our boards is to use signals with slower rise and fall times. This, too, is a desirable solution. Many people recommend using circuits with the slowest rise and fall times we can get away with to avoid reflection (and most other high-speed design) problems. The problem is (a) we may not have control over the rise time of the circuits we are using and (b) we may have already chosen the fastest rise time circuits available to meet other circuit requirements.

These are four legitimate solutions to our communications problem, but none of them may be practical for our specific situation. But there is one more:

---

[1] See Appendix G for information on how to download three audio files that illustrate this point.

5. We can acoustically engineer the gym to absorb reflections. Next time you are in a hotel or convention center conference room or hall, look closely at the surroundings. Chances are very good the room has been acoustically engineered to absorb echoes and reflections. It will have a soft carpet, cloth-paneled walls, and sound-absorbing ceiling tiles. Speakers are probably placed where there will be no phase shift between the sound coming from the presenter (wherever he or she is) and where you are sitting.

Similarly, we can (not acoustically, but) electrically engineer our trace to absorb reflections. Then there will be no reflections going back and forth on the trace to interfere with the receiver's ability to receive and understand the message the sender is sending.

The problem is this: In general, we don't know how to do that. We do know how to acoustically engineer a room, but we don't know how to electrically engineer a wire or trace to absorb reflections—with one very special exception. We do know what to do if we are dealing with what we call a *transmission line*.

TRANSMISSION LINES

So what makes transmission lines so special? Consider a long, straight wire or trace with its return wire or trace nearby. The wire has some inductance along its length. There is also some capacitive coupling between the wire and its return (see Figure 10-2). Figure 10-2 shows what we call a lumped model of the wire pair. It is called a lumped model because we show the capacitors and inductors as individual, lumped components. In reality, the inductance and capacitance are spread continuously along the wires. We don't know how to show that in a drawing, so we approximate it with a lumped approximation. On the other hand, this is fine because we are going to think of this wire as being infinitely long. Therefore, the part shown in Figure 10-2 is just an infinitely small part of the total length.

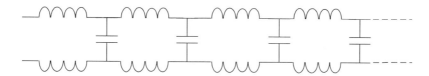

Figure 10-2 A transmission line is made up of an infinitely long network of capacitors and inductors.

Now, if these wires are infinitely long, there will be no reflection. At least if there is a reflection, it will take an infinitely long time for it to return, so we can assume it doesn't exist.

It takes at least one other thing to avoid reflections. The wires must be absolutely uniform. The little Ls and Cs must be identical everywhere along the wire. If the wires are not uniform, then we can consider them to be multiple sets of wires, each with different characteristics. Individual sets of wires will not be infinitely long, and therefore there will be reflections from each individual set. Therefore, the way to avoid reflections is to use an infinitely long, absolutely uniform wire or trace pair. We give such a wire or trace pair the special name *transmission line*.

It can be shown mathematically that if we look into the front of this wire pair there is an input impedance we can calculate. We give it the symbol Zo, and call it the *intrinsic impedance* of the line. If we could calculate the "lumped" values of inductance (L) and capacitance (C), the impedance would be calculated as shown in Equation 10-1:

$$Zo = \sqrt{L/C} \qquad [10\text{-}1]$$

Now here is a clever twist. If you look into the transmission line at the front, it looks like it has an impedance of Zo. Let's take our infinitely long transmission line and break it in two parts. If we look into the second part, it also looks like an infinitely long transmission line with an input impedance of Zo. What if we simply replaced the second part of the line with an impedance equal to Zo (see Figure 10-3)? From the front of the first line, it still looks exactly like an infinitely long line. It turns out it behaves that way, too. *A transmission line of finite length, terminated in its characteristic impedance, Zo, looks like an infinitely long transmission line.* Therefore, even though it has a finite length, it still will have no reflections. In this case it is not the infinite length that makes reflections irrelevant. It is the fact that the energy traveling down the line is exactly absorbed (dissipated) in the termination. There is no energy left to reflect back. The net effect is the same thing. There are no reflections to worry about.

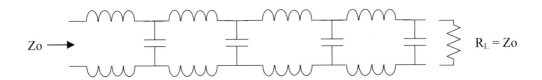

Figure 10-3 If we terminate a finite-length transmission line in its characteristic impedance, it looks infinitely long.

These are the two actions we need to take to control reflections on PCBs:

1. We need to make our traces look like transmission lines.
2. We need to terminate them in their characteristic impedances, Zo.

Of course, we can also make our traces short enough that reflections don't interfere with the receiver's ability to hear the signal in the first place, or we can slow down the rise time enough that the receiver can hear through the echoes.

There are certain types of transmission lines that are commonly used around us everyday. The coaxial cable leading to our cable TV is a 75-Ω transmission line. If you use 10Base2 coaxial cable for networking, that is a 50-Ω transmission line. The 300-Ω "twin lead" cable from your TV "rabbit ear" antenna to your TV is a transmission line. And it is no accident that those high-power electrical lines from the power generating plants to our cities are called transmission lines, strung along transmission line towers. Even power system engineers need to worry about reflections if the lines are long enough (about 480 miles or so).

The rest of this chapter is concerned with how to make our traces look like transmission lines, and what the implications are if we don't do that.

CRITICAL LENGTH

I have suggested that reflections (echoes) are not too bad a problem if the receiver is close enough to the sender. There is also less of a problem if the message is slowed down (the rise time is slowed down, or lengthened). If we think of a signal traveling down a trace, these two things are actually equivalent. The issue is this: What is the length of the signal path (in propagation time) relative to the rise (or fall) time of the signal?

For example, consider Figure 10-4. There are two cases shown, Driver A is driving a signal to Receiver B1, or to both Receivers B1 and to B2. B2 is much further away from the driver than B1.

The driver signal propagates down the trace. The horizontal axis represents both distance and time, because all signals on a particular trace layer (usually) propagate at the same, constant speed. Pick a spot (x) on the lower portion of the rising edge of the driver signal. Now move horizontally from that spot to the slope representing B1. This horizontal distance represents the propagation time between the driver and B1. Move horizontally an equal distance again, and then move up vertically to the driver slope (Point y). Notice that the driver is still driving at this point in time. We can think of the driver having momentum and overpowering the reflection, so the reflection has no effect on the circuit. (This is not an entirely accurate view of the dynamics going on here, but this helps to visualize what is happening.) B1 is "close" to the driver in this example.

On the other hand, move horizontally from the same point, x, to the point representing B2. This represents the propagation time to B2. Now, extend beyond B2 another similar increment in time, and then rise up vertically to the driver signal. By now, the driver signal has long since stabilized at its upper value, and the reflection may be significantly apparent on the trace, possibly disrupting the operation of the circuit. B2 is a "long" way away from the driver in this example.

The distance from the driver to B1 is intuitively short in this example, and the distance to B2 is intuitively long. Reflections have minimal impact on short traces and may have significant impact on long traces. How do we define the difference between a short trace and a long trace? The general rule of thumb accepted by most people goes like this:

- If the propagation time down the trace and back again is less than the rise time of the signal, then we can consider the trace to be short.

- If the propagation time down the trace and back again is longer than the rise time of the signal, then we must consider the trace to be long, and we must therefore consider whether terminations might be necessary.

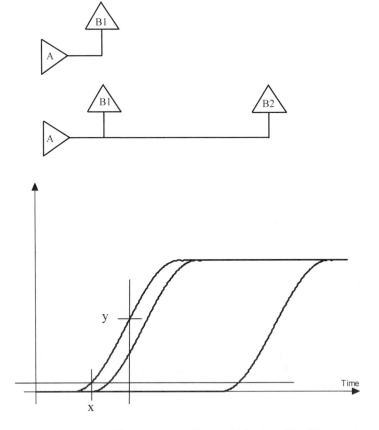

Figure 10-4 Receiver B1 is within the critical length, but B2 is beyond it.

Chapter 10

The boundary between these two lengths is called the *critical length* and is often defined as the length where the two-way delay of the line is less than the rise time of the pulse.

REFLECTION COEFFICIENTS

What causes reflections and what do they look like? If we construct our traces to look like transmission lines, we call them *controlled impedance* traces. We then should be able to control reflections. But if we don't control them correctly, reflections may still exist. The question is why and how large might they be.

If we have (tried to) use controlled impedance traces (transmission lines), at least two things might still cause a reflection in the system: (a) an impedance discontinuity along the line, and (b) an improper termination. Recall that the characteristic impedance of a transmission line is defined by its geometry (and the dielectric coefficient of the material surrounding the line). Any change in the geometry, or in the dielectric coefficient, will cause a change in the characteristic impedance of the line at that point. That may (and probably will) cause a reflection at that point. It is imperative that the geometry of our transmission lines be held constant everywhere along the line. (The implications of having lines change layers will be covered in a minute.) Also, if we want to control the characteristic impedance of a trace over several board fabrication runs, we need to fully specify (ideally in conjunction with the board fabricator) the materials to be used and their relative dielectric coefficient. If the geometry remains constant, but the materials change from run to run, the impedance will also change from run to run. At a minimum, this will require changing the termination values from run to run, something that might not even be possible.

Suppose we take as a given that the trace has been designed and constructed carefully with a constant and controlled characteristic impedance. If it is terminated correctly, there will be no reflection from the far end. But if the terminating resistor is not selected properly, there will be a reflection. The magnitude of the reflection is determined by a measure we call the *reflection coefficient*, ρ. In the case of a single, parallel terminating resistor at the far end of the trace, the voltage reflection coefficient is defined as shown in Equation 10-2:

$$\rho = \frac{RL - Zo}{RL + Zo} \qquad [10\text{-}2]$$

where RL is the terminating, or load, resistor.

Note that if RL = Zo, then the magnitude of the reflection coefficient goes to zero, and there is no reflection. If the trace is left unterminated (i.e., RL is infinite), then the reflection coefficient becomes RL/RL = 1. This means the reflection is 100% and it is positive in sign. So, if we send a 3-volt signal down an unterminated trace, the initial reflection back will be 3 volts for a total of 6 volts on the trace (at least at that moment).

Note also that if we short-circuit the trace, so that the terminating resistance is zero, then the reflection coefficient will become –Zo/Zo = –1. This means the reflection will be 100% but with the opposite sign. So, if we send the same 3-volt signal down a shorted trace, the initial reflection will be –3 volts, and the net signal on the trace will be zero volts (at least at that precise moment). (You might ask how you can put a voltage down a shorted trace. Remember that there is a propagation time before the signal reaches the far end. When the signal starts down the trace, it simply sees the characteristic impedance of the trace, Zo. It doesn't know the trace is shorted. It doesn't find that out until it reaches the far end. We're dealing with very short transient times here.)

The value of the reflection coefficient can range between –1 and +1. It is +1 for an open trace and –1 for a shorted trace. A reflection coefficient value of zero means the trace is "perfectly" terminated.

VISUALIZING REFLECTIONS

It can be difficult to visualize what reflections look like. One reason is because it depends on where you are looking at them. For example, do you want to see their impact along the trace, at the far end, or back at the near end? Do you want to look at them over time or at a single point in time? What does the basic signal look like before the reflections begin interfering with it? Are there multiple signals involved that are all interacting? UltraCAD has created a Transmission Line Simulator that you can download from the Web, described in Appendix F (Figure 10-5). This is one tool that can be used to help visualize the effects of various termination situations and problems.

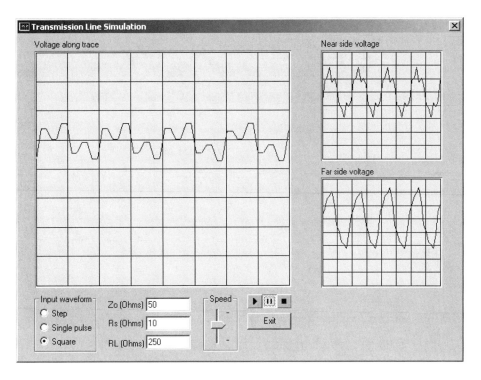

Figure 10-5 UltraCAD's Transmission Line Simulator, illustrating distortion along a trace cause by reflections. The driving input signal is a square wave.

Various commercial simulators exist to model transmission lines. One of them, Signal Vision, is used to illustrate various termination situations in the next chapter. Figures 10-6 and 10-7 illustrate simulations of terminations that are too low and too high, from both the Signal Vision tool and from the Hyperlynx LineSim tool. Both tools are products of Mentor Graphics.

DETERMINING TRACE IMPEDANCE

With the exception of the effect of the material's relative dielectric coefficient (ε_r), the characteristic impedance of a trace is determined purely by its geometry. That means the board designer has complete control over it (at least in theory). The formulas for determining and defining impedance are very complex. The

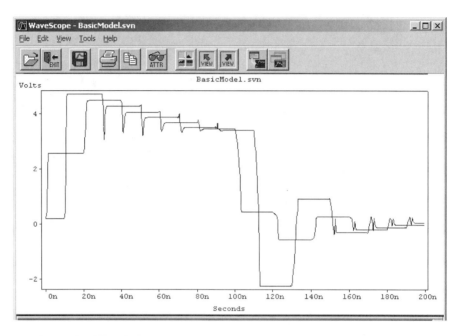

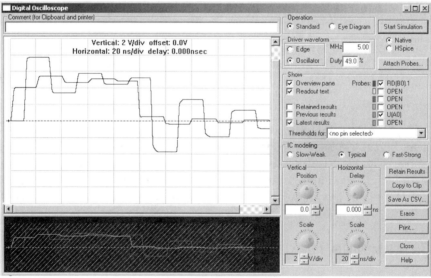

Figure 10-6 Simulation of a driver driving a 50-Ω transmission line with 1,000-Ω termination. Upper illustration is from the Signal Vision tool, lower one from the Hyperlynx LineSim tool.

Chapter 10

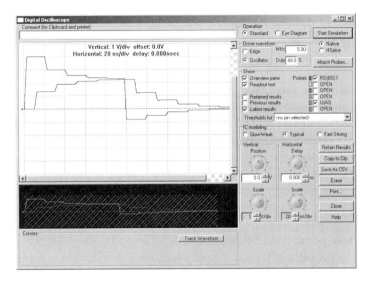

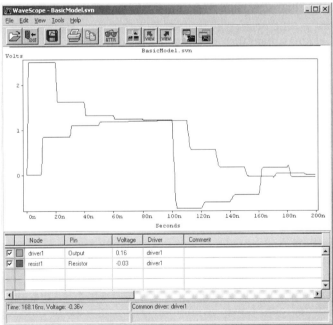

Figure 10-7 Simulation of a driver driving a 50-Ω transmission line with 10-Ω termination. Upper illustration is from the Signal Vision tool, lower one from the Hyperlynx LineSim tool.

primary ones are shown in the Endnote at the end of this chapter. These formulas can be very difficult to work with and therefore are not particularly useful.

Fortunately, many alternatives have become available for designers to use for calculating the impedance of their traces. UltraCAD has made available a freeware calculator for doing these calculations (Figure 10-8). Refer to Appendix H for information about UltraCAD's freeware calculators. Polar Instruments has a particularly good commercially available calculator available in several versions from their Web site (also shown in Figure 10-8). Many of the higher-end PCB design packages include built-in trace impedance calculators. Numerous other Web-based or Web-obtainable design aids have been developed in the last few years. Do a search on the Web for "PCB trace impedance" or variations on this.

Both propagation time and characteristic impedance can be impacted by capacitive loading along the trace. Every receiving device on the trace adds a small increment of lumped capacitance to the trace. Even vias can contribute small increments of capacitance.

The adjustment factor commonly used is shown in Equation 10-3:

$$\sqrt{1 + \frac{Cd}{Co \times l}} \qquad [10\text{-}3]$$

where Cd is the total amount of all such capacitances added, Co is the characteristic capacitance of the line, per unit length, and l is the length of the trace. The application of this adjustment factor to characteristic impedance and to propagation delay is shown in the Chapter Endnote.

Despite any claims to the contrary, impedance calculations are by their very nature approximate. The calculations are based on approximations that are reasonable, but cannot be exact. That is because, in part, the nature of the physics involved is just too complex to model exactly without any error. Therefore, it is expected that calculations based on tools from different sources might reach slightly different results. The better tools, however, should agree within about 2%.

Your board fabricator, however, will usually be doing well to hit the impedance target within about 5% because of the many variables (times, temperatures, pressures, material dimensions, and even humidity) that exist in the chemically based board fabrication process itself. If you want impedance control tighter than

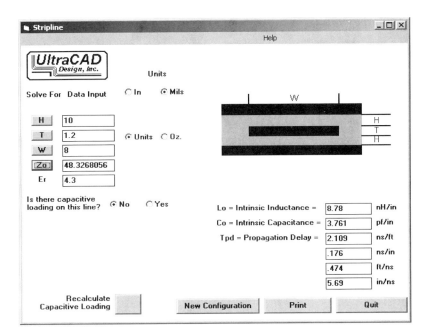

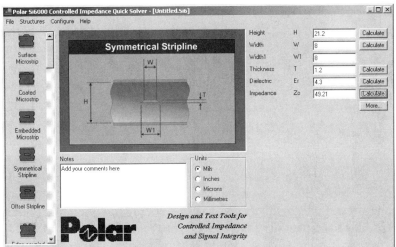

Figure 10-8 Two impedance calculators: UltraCAD's freeware calculator, UltraCLC, and Polar Instruments SI6000 QuickSolver, both available from their respective Web sites.

about 5%, your selection of board fabricators that can do the job will shrink significantly and the prices will start going up.

Not only do these tolerances cause impedance to vary from board to board, they will cause impedance variations from place to place on the same board. In 1994 UltraCAD partnered in a study to determine the effects of vias on traces. An unexpected by-product of that study was information regarding the variation of impedance between boards fabricated in the same batch process and even along traces on the same board. Figure 10-9 is taken from that study. It shows a TDR pattern of one of the traces on a test board. The trace is 15 inches long and has three vias along its length, but the trace does not pass through the via. Instead, the trace stays on the same layer for the entire 15 inches. The impedance varies by about 3% along this trace.[2]

There are several reasons for this kind of result along a trace: temperature gradients during fab, pressure gradients during fab, nonuniform thicknesses and etching, material nonlinearities, and so on. Polar Instruments has done some interesting work to show that another effect can also come into play. One of the most common board materials used in board fabrication is FR4. FR4 is a blend of resin and glass, but it is not a uniform, homogeneous blend. The glass tends to form in clumps and the resin fills in around it (see Figure 10-10). The glass has an effective ε_r of approximately 6.0 and the resin an ε_r of approximately 3.0. These combine, on average, to yield an effective ε_r of approximately 4.0 to 4.3. But the ε_r can vary dramatically from this at any specific point, depending on the specific mixture of glass and resin at that point. Since the effective ε_r can change so dramatically from point to point, so can the impedance that depends on that ε_r.

TERMINATION TECHNIQUES

It is one thing to know that an impedance-controlled trace needs to be terminated correctly. It is quite another thing to know how to do that. It is important to note, however, that it is not the PCB designer's job to design the termination scheme. It is the circuit design engineer's job to do that. The board designer's job is to know whether or not a termination scheme is being used at all, and to raise the question if it appears that a termination scheme has been inadvertently overlooked.

---

[2]See "The Effects of Vias on PCB Traces," 1994, available from UltraCAD's Web site. This article also appeared in *Printed Circuit Design* magazine in August 1996. Also see Appendix I for information on how TDRs measure the impedance of traces.

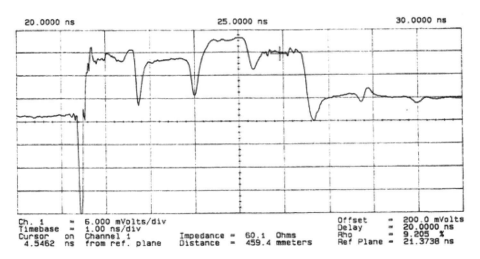

Figure 10-9 TDR results of a PCB trace illustrating impedance variations at different points along the trace (courtesy UltraCAD Design, Inc.).

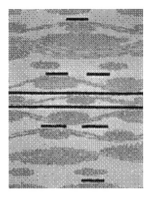

Figure 10-10 FR4 is made up of glass and resin. The apparent combination of these can vary from point to point on the board. This is a copy of a microsection cut from a test board (courtesy Polar Instruments).

I can tell you from personal experience that everyone wins when you raise the question about terminations. If no change is necessary, you look intelligent simply for knowing about transmission lines, impedances, and terminations. If a change is necessary, you become a hero for being the one to call it to your engineer's or customer's attention. Either way, you gain in stature.

In this chapter we discuss the five most common forms of transmission line terminations.[3]

Parallel Parallel termination (Figure 10-11) is the most intuitive to understand. A single resistor equal to the characteristic impedance of the line (RL = Zo) is attached to the end of the line. All the energy flowing down the trace is absorbed in the resistor and there is no reflection. The terminating resistor can be connected to either Vcc or ground.

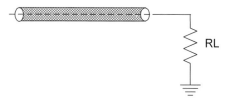

Figure 10-11 Parallel termination.

This technique has several advantages: (a) the value of the resistor is relatively easy to determine, (b) there is only a single component, (c) it is easily connected, and (d) it performs well with distributed loads (i.e., loads that are distributed along the trace). There is only one drawback to this type of termination: It provides a continuous DC path to ground. Therefore, continuous DC current can flow through it at all times. This may or may not be an issue for a single trace. But if your design has 1,000 or so impedance-controlled nets, the total power dissipation in 1,000 terminating resistors can become quite significant.

Thevenin A closely related variation of the parallel termination strategy is the Thevenin termination (Figure 10-12). This consists of a pair of resistors, one going to ground and one going to Vcc. The pair of resistors provides a pull-up/pull-down function as well as a termination function. Therefore, it can improve noise margins in certain situations and performs as well as the parallel termination with distributed loads.

---

[3]For a good discussion of trace terminating techniques see Ethirajan and Nemec, "Termination Techniques for High Speed Buses," EDN, February 16, 1998, p. 135.

Chapter 10

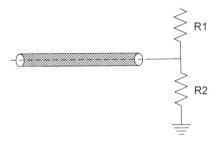

Figure 10-12 Thevenin termination.

On the one hand, it would appear that the selection of resistor values would be relatively straightforward. The parallel combination of the two resistors must simply equal Zo, or (c.f. Equation 5-9):

$$Zo = \frac{1}{\frac{1}{R1} + \frac{1}{R2}}$$

But Ethirajan and Nemec showed that there are optimum values for R1 and R2 based on the specific characteristics of the driver. The optimum values for R1 and R2 primarily optimize (minimize) the power dissipated in the circuit. This optimum value can be difficult to determine. Other drawbacks to this strategy include the addition of an additional component, and the fact that DC current still flows through the resistor pair at all times. Finally, this strategy is only well-suited for bipolar (two-state) devices, not for tristate logic families.

Some people have reported that there can be an EMI problem with this termination strategy. Note that there are different currents flowing between Vcc and ground, through R1 and R2, depending on the logic state. This current changes at the same rate (di/dt) as the logic state changes (the rise/fall time of the signal). In this respect, the two resistors look exactly like a switching logic gate. Therefore, they might need to be decoupled (with bypass capacitors) just as a logic gate might need to be decoupled (with bypass capacitors). The issue really is related to loop areas and currents, which are one of the primary causes of EMI. Thus, you may occasionally hear people say that Thevenin terminations may also require decoupling capacitors to quiet down EMI emissions.

AC Termination Yet another variation is the addition of a capacitor in series with the parallel terminating resistor (Figure 10-13). The primary advantage of this is that the capacitor blocks DC current, so there is no steady-state current flowing through the termination. At first glance it would appear that this strategy otherwise has all the advantages of the parallel termination strategy. However, the "cost" of this, of course, is the added component.

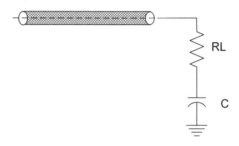

Figure 10-13 AC termination.

There are some other subtle and some not-so-subtle problems with AC termination. If a large capacitor value is used, there can be considerable power dissipation, and the strategy is little different from normal parallel termination. If a very small capacitor is used, it will cause overshoot and may interfere with the rise and fall times of the signal.

There are also some subtle interactions that take place with AC termination using a small capacitor. As the voltage across the terminating resistor changes, the current through the capacitor changes. This causes the capacitor to charge or discharge with an RC time constant related to the component values. If the time constant is too short, the capacitor charges during the half-cycle and the voltage at the receiver changes (and distorts) accordingly. Very long lines can also lead to signal distortion.[4]

Series Series termination (Figure 10-14) is becoming more and more common in today's high-speed designs. It has the two desirable attributes of a single component and no DC current draw at all. The series termination resistor is placed at the front of the trace, however, not the end; the end is left open-circuited. There-

---

[4]Brooks, Douglas, "Transmission Line Terminations, It's the End That Counts!" available from Mentor's Web site at www.mentor.com/pcb/tech_papers.cfm.

fore, there is a 100% positive reflection from the far end of the trace that reflects back toward the front.

Figure 10-14 Series termination.

The fact that there is a single reflection traveling back up the line is of little consequence if there are no devices along the line; that is, if the loads are lumped at the far end of the line. But if the loads are distributed along the line, the loads will see different signal (voltage) levels depending on which way the signal is traveling. Some designs actually take advantage of this and are designed to switch only at the higher signal level. The computer PCI bus is one example of this.

The value for the series termination resistor is set so that the sum of it and the output impedance of the driver totals to the impedance of the trace. That way, there is no secondary reflection from the driver. The voltage pattern we expect to see is a single, positive reflection from the far end of the trace that travels to the front of the trace and is completely absorbed in the source resistor and driver.

That means that we have to know the output impedance of the driver. That isn't too serious a problem, since the output impedance should be specified by the manufacturer. But an additional complication is that the output impedance is often *different* depending on the output state. That is, the output impedance of the logical high state may be different from the output impedance of the logical low state, making it impossible to properly terminate both states with a single series resistor.

Diode The diode termination technique is not designed to minimize reflections, only to limit them. Two diodes are place on the line as shown in Figure 10-15. Reflections can exist at both ends of the line and will travel back and forth along the line. This technique simply limits the magnitude of the reflection to between Vcc and ground (or more properly, one diode drop above or below these levels, respectively). There is no power impact of this technique, and the diodes can be placed anywhere along the line. In high-speed designs, the diodes must be very fast switching devices.

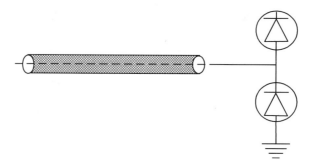

Figure 10-15 Diode termination.

SOME DESIGN ISSUES

In this section we discuss a few design issues related to controlled impedance traces that can be very important.

Changing Trace Layers 1 We have indicated that for there to be impedance control, and therefore control over reflections, the impedance of the trace must be constant over its entire length. Since impedance is a function of geometry, this implies the geometry must be constant, also. Consider, however, what happens when a trace changes layer (Figure 10-16). It is important that the *impedance* of the trace stays constant. But the geometry has changed, the relationship to the planes has changed, and the (relative dielectric constant of the) material may have changed. Therefore, it is highly likely that if we keep the same trace width, the impedance will change.

Since the environment often changes when we change layers, we usually have to also adjust the trace width when we change layers. This is necessary because what is important is maintaining a constant impedance (not just a constant geometry) on each layer. If we are designing with a certain impedance in mind (say 50 Ω) we have to determine what a 50-Ω trace looks like on *each layer* of our board. Then, if we move a 50-Ω trace to that layer, we may have to change the trace's properties to maintain the desired impedance on that particular layer.

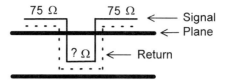

Figure 10-16 When changing layers, care must be taken to ensure that the impedance of the trace is constant on all layers, even if that means the geometry needs to change.

Power Planes The location of the return signal is an integral part of the geometry of impedance determination and control. This point simply cannot be overemphasized. For coaxial and twin lead cables, the location of the return current is fixed by the design of the cable. On PCBs, the return signal wants to return directly underneath the trace (refer to Appendix B). This simple concept has some significant implications.

All signals have a return signal associated with them. Every single one. You (as the PCB designer) need to know where they are. It is ironic that some designers spend an enormous amount of time worrying about, strategizing about, and routing 50% of the signals. They then ignore the other 50% of the signals (the return signals) assuming they will take care of themselves.

It is (almost) impossible to achieve impedance control on a PCB without a reference plane for the return signal to travel on. Thus, unless you are really good, and have some really well-thought-out strategies, all high-speed designs need power and ground planes. It is (almost) impossible to achieve impedance control without them.

The return signals can, and do, travel on either the power or the ground plane. Power and ground planes are the same thing to AC signals. The distinction is only a DC distinction. (If you have trouble visualizing this, consider how many capacitors—which pass AC currents—there are between the power and ground planes.) Typically, the return signal will be on the closest plane to the signal trace, which we then call the *reference* plane.

Changing Trace Layers 2 If we do anything to prevent the return signal from following directly underneath the trace, we impact the geometry. This, then, impacts the magnitude of the impedance, causes an impedance discontinuity, and therefore causes a possible reflection. Therefore, we must be careful in our designs to never do anything that restricts the return signal.

Figure 10-17 illustrates three typical routing situations. (Note that this is the same figure as Figure 9-10 in Chapter 9. In Chapter 9 we talked about this issue from the standpoint of loop areas and EMI.) In Figure 10-17a, a trace is routed above a reference plane and the return signal (dotted line) returns on the plane directly under the trace. In Figure 10-17b, the trace routes through a via and proceeds on the trace layer on the other side of the reference plane. The return signal crosses over to the other side of the plane, and remains directly under the trace. In neither case are we interrupting the return signal path. There is no impedance discontinuity in either of these situations (assuming we have constructed the trace itself correctly on each layer).

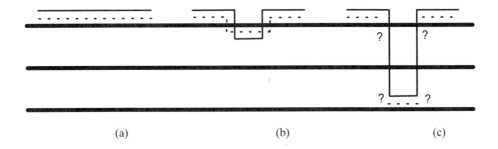

Figure 10-17 It is important to know where the return currents are to maintain impedance control.

What about Figure 10-17c? Here the trace routes through a via to a completely different trace layer. To maintain impedance continuity, the return signal must find its way to a new reference plane several layers away. A good question is this: How does it do that? Unfortunately there is not a really satisfying answer.

We know this situation works because lots of people (including us) do this. The practical answer is that the return signal travels to the other plane through a nearby bypass capacitor (since there are lots of them distributed around our board). But it is possible that the path to the bypass capacitor and back under the trace creates an impedance discontinuity that may then create an unwanted reflection.

Different engineers have different attitudes toward this:

1. Some ignore the problem as being relatively minor.
2. Some (e.g., Henry Ott and Howard Johnson in some e-mail forums) have suggested we might consider placing bypass capacitors beside each via for the specific purpose of providing a path for the return signal (for both impedance control reasons and EMI loop area reasons).
3. A few take the position that no traces at all shall be routed in this manner to ensure that this type of reflection problem cannot happen.

Most people consider the third option to be extreme. But if your engineer takes this position, at least you now understand what his or her thinking is. After all, the engineer is the customer and we do what the engineer asks.

Note: If you want to know what UltraCAD does, we follow the first option, and so far have not experienced a problem. If you fell into the trap of considering a "hard" or solid via between the two planes, consider the consequences if one plane is ground and the other is Vcc. You can end up with a lot of angry electrons this way, at least for a moment.

Slots in Planes 2 As we have emphasized, the geometry associated with the traces must be controlled. This means, in particular, that the return signal must be allowed to follow directly on the plane underneath the trace. Consider a situation illustrated in Figure 10-18. (Note that this is the same figure as Figure 9-7. In Chapter 9 we said this could cause an EMI problem. You will see this figure yet again when we talk about crosstalk. There are several reasons not to allow slots on planes.) A slot exists, for some reason, on the plane. A controlled-impedance trace routes across the slot. The return signal must loop up around the slot and back again to the trace. The geometry of the trace has changed at this point, causing an impedance discontinuity. This is analogous to cutting a coaxial cable and then resplicing it as suggested in the figure. If you have ever done this with your TV cable, you ended up with a severe ghosting problem on the TV caused by the reflections, which were in turn caused by the impedance discontinuity at the splice.

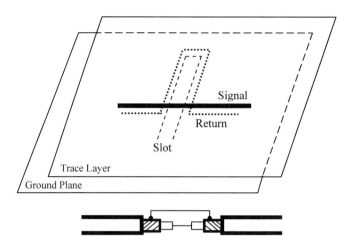

Figure 10-18 A discontinuity on the plane can cause the return signal to loop, causing, in turn, an impedance discontinuity.

STUBS

In the last chapter we talked about stubs from an EMI standpoint. There is also an impedance aspect to stubs that is important. Figure 10-19 (the same as Figure 9-12 from the previous chapter) illustrates how stubs can exist when we make a connection from a trace to a pad or device. The stubs are unterminated sections radiating away from a controlled impedance trace, almost like the "Y" problem we will talk about in the next chapter. Signals can reflect from the far (unterminated) end of the stub and cause signal integrity problems.

Stubs S1 and S3 are short enough (well within the critical length) so as not to cause much of a problem. Stub S2, however, will result in a significant impedance discontinuity and damaging reflections, and such long stubs are to be avoided. Obviously, Stub S4 will cause a reflection problem.

Chapter 10

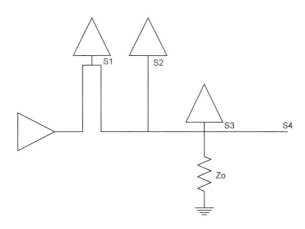

Figure 10-19 Stubs are unterminated transmission line segments and must be kept short to avoid reflections.

The rule for stub length most people use is about 1/10 or 1/12 the rise time. For a 1 ns rise time (about 6 inches) the rule would be about 75 to 100 ps, or about .5 inche. I have run simulations with signal integrity models and found that reflections are just becoming detectable in this range. Nevertheless, good design practice would be to keep stub lengths shorter than this. Lengths less than .25 inches are usually easily achievable.

ABSOLUTE VS. RELATIVE VALUE OF ZO

A common question board designers ask is "What should the value of Zo be? What value should I use?" With a few exceptions, the answer is that it doesn't matter much. The value of Zo is not nearly as important as is the fact that (whatever it is) (a) it is constant everywhere along the trace, and (b) that it is terminated properly with that value (whatever it is).

The absolute value of Zo is important under these conditions (there may be other special cases that should be obvious when they come up):

1. If the trace is mating to the outside world, usually through an impedance-controlled connector, then the trace must match the connector.

2. If the trace is mating with a trace on another board, again probably through a connector, a specified impedance is probably necessary because all boards must "play" together without special matching.
3. A very few special components have strict impedance-matching requirements.
4. The circuit designer may specify an impedance level because it is easier to select or specify termination resistors or circuits for the board. (Note that this is not a technical requirement, but is a policy requirement.)

Otherwise, devices are reasonably tolerant of the impedance of the trace, and a specific value of trace impedance is not particularly important.

We do a lot of boards in the video processing industry. The signals frequently come onto and leave the board through 75-Ω connectors. Obviously the traces leading to and from these connectors must also be 75-Ω traces. But the traces on the inner circuits don't need to be 75 Ω. Nevertheless, some of our customers require 75-Ω traces everywhere on the board, because "these are video signals." It's as if there is something special about a video waveform that only works on a 75-Ω trace.

There is not uniform agreement among engineers regarding whether we should use relatively higher trace impedances (say 80 Ω or more) or relatively lower trace impedances (say 30 Ω or so). If you subscribe to various e-mail forums you will hear different theories about either of these extremes, as well as why we should all stick with 50 Ω.

One argument says that higher impedance results in lower currents and therefore lower EMI problems. Others argue that there are offsetting voltage issues with higher impedances that negate any EMI gains. Another argument is that higher impedance creates more voltage-related noise problems on the board. Still others argue that lower impedances cause increased current and therefore power supply requirements. Still others try to offer proof that 50 Ω offers the best overall compromise. I have not heard any convincing argument that there is any impedance level that is inherently better than any other within the reasonable range of about 30 Ω to about 100 Ω.

Chapter Endnote

ON FORMULAS

These are the generally accepted formulas for impedance-related calculations and can be found, except where noted, in several publications.[5]

In the following formulas:
    Dimensions are in mils
    ε_r = Relative dielectric coefficient
    ε = base of the natural logarithm ($\cong 2.71828$)
    Ln = natural logarithm.

Microstrip:

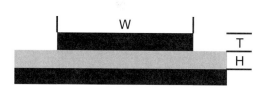

$$Zo = \frac{87}{\sqrt{\varepsilon_r + 1.41}} Ln\left(\frac{5.98H}{.8W + T}\right)$$

---

[5]IPC-D-317A, "Design Guidelines for Electronic Packaging Utilizing High-Speed Techniques," January 1995, IPC p. 22ff.

$$Co = \frac{.67(\varepsilon_r + 1.41)}{Ln\left[5.98H / (.8W + T)\right]}$$

Stripline:

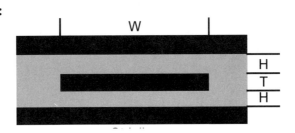

$$Zo = \frac{60}{\sqrt{\varepsilon_r}} Ln\left(\frac{1.9(2H + T)}{.8W + T}\right)$$

$$Co = \frac{1.41\varepsilon_r}{Ln\left[3.81H / (.8W + T)\right]}$$

Dual Stripline:

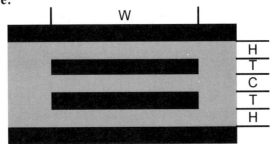

$$Zo = \frac{80}{\sqrt{\varepsilon_r}} Ln\left[\frac{1.9(2H + T)}{.8W + T}\right]\left[1 - \frac{H}{4(H + C + T)}\right]$$

$$Co = \frac{2.82\varepsilon_r}{Ln\left[\frac{2(H-T)}{(.268W + .335T)}\right]}$$

Embedded Microstrip:[6]

$$Zo = \frac{60}{\sqrt{\varepsilon_r\left[1 - \varepsilon^{\left(-1.55H_1/H\right)}\right]}} Ln\left(\frac{5.98H}{.8W + T}\right)$$

Propagation Time (nanoseconds/inch):

Stripline:

$$\text{PropagationTime}(T_{pd}) = \frac{\sqrt{\varepsilon_r}}{11.8}$$

Microstrip:[7]

$$\text{PropagationTime}(T_{pd}) = Br \times \frac{\sqrt{\varepsilon_r}}{11.8}$$

Where:
$Br = .8566 + .0294Ln(W) - .00239H - .0101\varepsilon_r$
W = trace width (mils)
H = distance between the trace and the plane (mils)
ε_r = relative dielectric material between the trace and the underlying plane
Ln represents the natural logarithm, base e

---

[6] Brooks, Douglas, "Embedded Microstrip Impedance Formula," *Printed Circuit Design* magazine, February 2000.
[7] See Chapter 2.

Corrections for capacitive loading:

The commonly used correction factor for capacitive loading of a trace is

$$\sqrt{1 + \frac{C_d}{C_o \times l}}$$

where C_d = capacitance added by the device loads
  C_o = intrinsic capacitance of the line per unit length
  l = length of the line

Its proper usage for this correction factor for impedance (Zo) and propagation times (Tpd) is:

$$Z_o' = \frac{Z_o}{\sqrt{1 + \frac{C_d}{C_o \times l}}}$$

$$T_{pd}' = T_{pd}\sqrt{1 + \frac{C_d}{C_o \times l}}$$

11 SOME TRANSMISSION LINE SIMULATIONS

*M*entor Graphic's Signal Vision simulator is a useful tool for illustrating transmission line issues. In this chapter we look at several simulations of both good and bad designs to see what the differences are between them.

BASIC SIMULATION

Figure 11-1 illustrates a basic simulation setup. The individual elements of the simulation are selected and placed on the work surface and connected together. The elements are then given values for analysis.

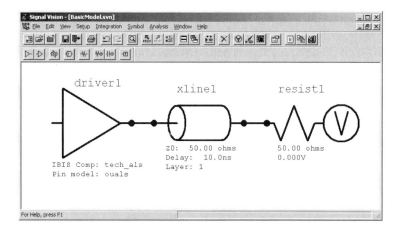

Figure 11-1 Basic simulator setup.

Figure 11-2 shows the various screens for defining the values. It is almost always necessary with any manufacturer's simulation tool to define a software model for any active device used, in this case the driver. These software models are usually (but not always) IBIS models.

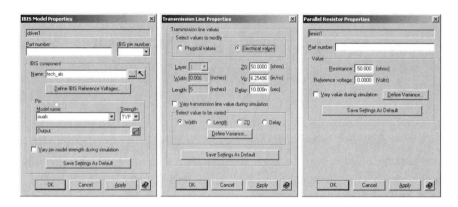

Figure 11-2 Submenus for setting the element parameters for the simulation.

Double-clicking the driver element produces the first submenu shown in Figure 11-2. We have selected a generic IBIS model provided by Mentor Graphics, designated as "tech_als." We select this model because it is convenient and because it has a fast rise time (approximately 1.0 ns).

Double-clicking the transmission line element brings up the second submenu. We can define the transmission line in terms of its board stackup parameters or by its electrical parameters. The board stackup parameters have obvious value if we are investigating real board parameters or if we are integrating the Signal Vision tool with the Expedition board design tool and analyzing an actual design in progress. In this instance we select electrical parameters to define a 50-Ω line 10 ns long.

Finally, we double-click the load resistor and select the loading for the line, shown in the third submenu. Here we have selected 50 Ω for our initial simulation, tied to zero volts (ground).

We could add other elements to the simulation and change any of the parameters through these various screens.

Finally, we set up the simulator control, Figure 11-3. Here we are telling the simulator to cover a time period of 20 ns with a pulse that is starting low and rises, and stays "high" for 19 ns. That is, we a simulating a step function rise in signal and looking at it for 19 ns.

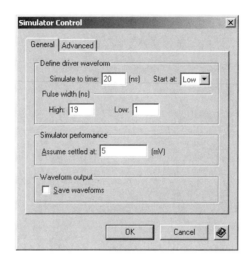

Figure 11-3 Simulator control submenu.

The result is shown in Figure 11-4 and looks just like we would expect for a properly terminated transmission line. The driver signal immediately rises to 2.5 volts and 10 ns later the voltage at the load resistor rises to 2.5 volts. This time delay is the propagation time down the 10 ns transmission line. Since the transmission line is terminated correctly, there are no reflections.

Now let's do something more interesting. We'll terminate the line with a 150-Ω resistor and see what the results are. To see the results clearly, we'll need to extend the simulation time much further out, in this case to 100 ns. The result of the simulation is shown in Figure 11-5.

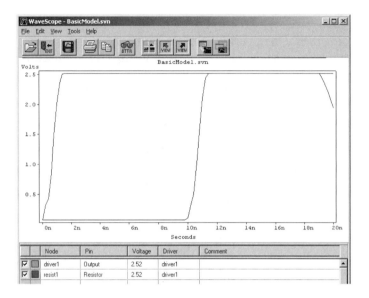

Figure 11-4 Simulator response for a properly terminated line.

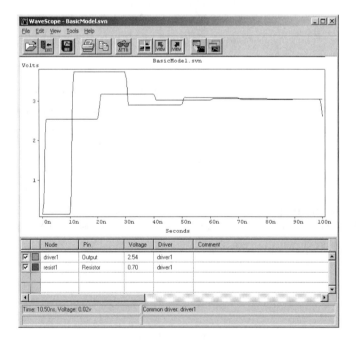

Figure 11-5 Simulator response for a 150-Ω termination.

To fully appreciate what is happening here, we need to look at the equivalent circuit of the simulation, including the Thevenin equivalent circuit of the driver in its "high" state (see Chapter 5). The tech_als software driver model used here has an output impedance of about 18 Ω in the "high" state (Figure 11-6). The trick is, the equivalent circuit for the simulation is different for the very first instance than it is in the steady state. In the very first instance (fast rise time), the transmission line looks (to the driver) like a 50-Ω load (a). The driver has not yet seen the load (whatever value it is) at the far end of the trace. In the steady (low-frequency) state the driver sees *only* the load at the end of the line (b) and no longer the transmission line. Figure 11-5 is a representation of what happens as we transition between these two times (initial response and long-term stabilization).

Figure 11-6 shows the equivalent circuit for this simulation. At the very first moment, the driver sees the 50-Ω load of the transmission line. The 50-Ω transmission line and the 18-Ω source impedance of the driver combine to form a voltage divider (see Chapter 5). The voltage at the driver output immediately rises to (50/68)3.4 = 2.5 volts. This 2.5-volt signal propagates down the transmission line to the load and arrives at the load 10 ns later.

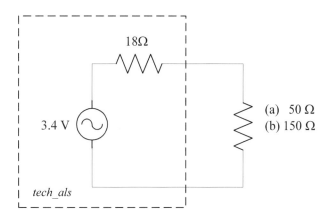

Figurer 11-6 Thevenin equivalent circuit for the driver.
The transient load (a) the driver sees is different from
the steady state load (b).

Because the line is not terminated correctly, the signal reflects off the load with a positive reflection coefficient (see previous chapter) of

$$\rho = (RL - Zo)/(RL + Zo) = (150 - 50)/(150 + 50) = .50$$

A reflection of .5 x 2.5 = 1.25 volts occurs, so the reflected signal traveling back up the line toward the driver is the initial 2.5 volts plus the 1.25-volt reflection, or 3.75 volts. Note that this is higher than the maximum output available from the driver itself (3.4 volts).

Ten ns later (at 20 ns) this reflected signal arrives back at the driver. But now there is a negative reflection because the 50-Ω transmission line sees a source impedance (at the driver) of only 18 Ω. At the source end the reflection coefficient is

$$\rho = (RL - Zo)/(RL + Zo) = (18 - 50)/(18 + 50) = -.47$$

The 1.25-volt reflected signal is reflected again, this time with a magnitude of $-.47$ x 1.25 = $-.59$ volts. The voltage at the driver is now 3.75 minus this reflected .59 volts or 3.20 volts. The reflected signal continues back and forth in this manner until it dies out and stabilizes at its long-term (DC) value.

The long-term DC value is found from Figure 11-6 using the long-term load value of 150 Ω. The long-term stable value is (150/168)3.4 = 3.0 volts, which can also be seen from simulation, Figure 11-5.[1]

Now let's look at the same model with only a 20-Ω load resistor. The model result is shown in Figure 11-7. As before, the voltage at the driver output immediately rises to (50/68)3.4 = 2.5 volts. When that 2.5-volt signal arrives at the far end of the transmission line, it reflects with a negative reflection coefficient equal to

$$\rho = (RL - Zo)/(RL + Zo) = (20 - 50)/(20 + 50) = -.43$$

The magnitude of the reflected signal is $-.43$ x 2.5 = -1.07 volts. So the immediate signal at the load is 2.5 − 1.07 = 1.43 volts. This signal travels back toward the driver, where the reflection coefficient is $-.47$, as before. So the magnitude of the next reflection is .47 x 1.07 = .50 volts. Therefore, the signal level at the driver output just after 20 ns is 2.5 − 1.07 − (−.5) = 1.93 volts.

---

[1] This kind of problem is solved manually through a lattice diagram. There are numerous sources, including the Web, for information on using lattice diagrams to solve reflection problems.

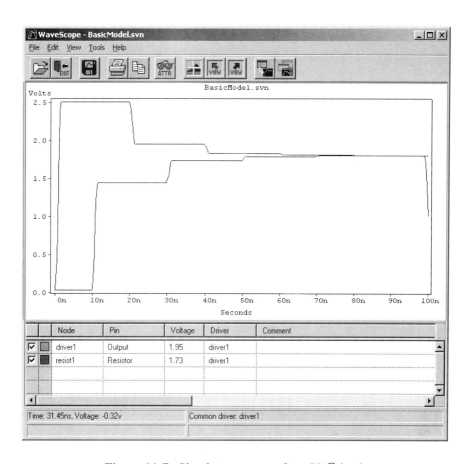

Figure 11-7 Simulator response for a 20-Ω load.

As before, the reflections continue until they die out. The system will stabilize at a DC value. The 20-Ω load combined with the 18-Ω source impedance yields a (20/38)3.4 = 1.79 final voltage.

As one final basic illustration, let's select a perhaps more realistic simulation. We'll set the trace length at 2 ns (equal to about 6 in) and simulate a square pulse width of 12 ns high and then 12 ns low. If we perfectly terminated the line, we would expect a perfect square wave at the load end on the line. Each waveform would be stable at either 0 volts or 2.5 volts, and the load waveform would follow the driver waveform by 2 ns. Figure 11-8 illustrates the case of a poorly terminated line (150 Ω). As can be seen from the figure, the waveforms are not very clean and could very well cause system logic errors because of the incorrect terminations.

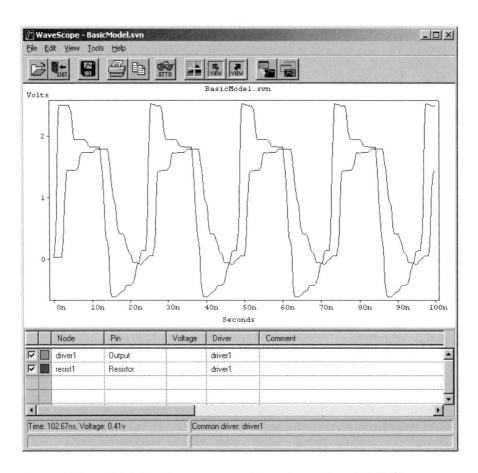

Figure 11-8 Simulator response of a poorly terminated (150-Ω) line to a repetitive square wave input signal.

Chapter 11

SERIES TERMINATION

As described in the previous chapter, series termination allows for a full reflection at the far end of the line. The reflection is absorbed at the source end, however, and no further reflections exist. Figure 11-9 is an equivalent circuit for our simulation of series termination.

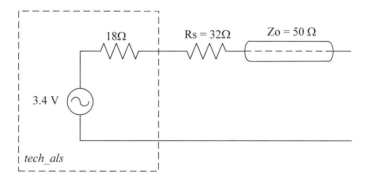

Figure 11-9 Equivalent circuit for a series termination.

The series terminating resistor, Rs, is chosen so that the sum of it and the source impedance of the (tech_als) driver equal the Zo of the line (50 Ω). Figure 11-10 is our Signal Vision model for simulating this circuit. The far-end 10k resistor is used to simulate an unterminated line yet make provision for a measuring point to display on the model output. It has negligible impact on the actual simulation.

The result of the simulation is shown in Figure 11-11. The two curves shown in Figure 11-11 are for the signals at the driver output and at the far end of the 50-Ω trace. The source resistance of the driver and the series terminating resistor together form a voltage divider with the 50-Ω trace. So the voltage at the near end of the trace rises to one-half the 3.4-volt source voltage. Two ns later this 1.7-volt signal arrives at the far (unterminated) end of the line where the reflection coefficient is 100%. The reflected signal is then 3.4 volts. When this reflected signal arrives back at the source resistor, the

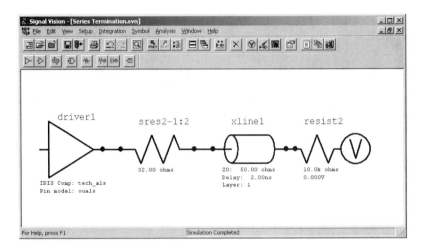

Figure 11-10 Signal Vision model for series termination.

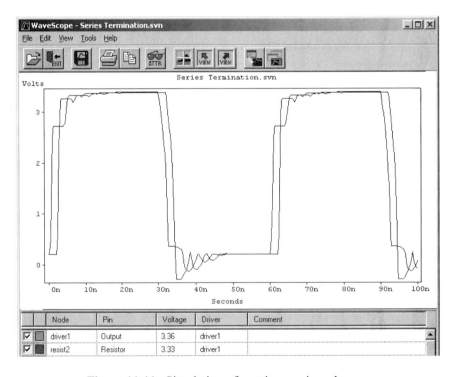

Figure 11-11 Simulation of a series terminated trace.

energy is absorbed and the signal stabilizes at the 3.4-volt source level without further reflections.[2]

PLACEMENT ISSUES

Figure 11-12a shows the ideal termination placement for a parallel termination. Figure 11-12b shows a resulting simulation. The results seem pretty clean. But what if we have to place the terminating resistor perhaps an inch in front of the receiver? (Note: Admittedly an inch might be a worst-case scenario, and there might be some other solutions available, such as internal routing channels and populating the back side of the board. Nevertheless, we sometimes are faced with dimensions of this magnitude.)

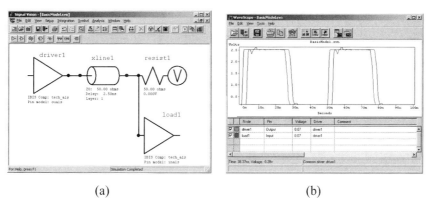

(a)   (b)

Figure 11-12 Ideal placement of a parallel termination.

If we place the terminating resistor an inch in front of the receiver, that means there is an unterminated transmission line segment between the terminating resistor and the receiver. The simulation for this, and its result, are shown in Figure 11-13a and 11-13b. There are some pretty healthy spikes that result from this placement.

---

[2]The signals are not perfectly clean in the simulation because the source impedance in the model is not perfectly linear, nor is the source resistance the same for both the high and low signal states.

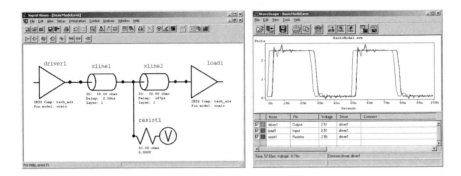

Figure 11-13 Placing the terminating resistor before the final device can cause reflections.

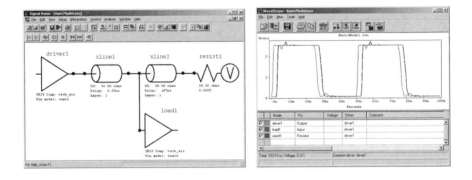

Figure 11-14 Placing the termination resistor after the final device is nearly identical to the ideal placement.

A better solution is simple but the implications are not always recognized by designers. Instead of placing the termination resistor *before* the receiver, place it *after* the receiver. Seems simple, doesn't it? The difference is that now there is a terminated transmission line segment between the device and the termination. Before, there was an unterminated segment; now there is no unterminated segment. Figure 11-14a and 11-14b show this simulation and its result. The results are virtually identical to Figure 11-12a and Figure 1-12b.

The simple act of placing the termination after the receiver instead of before the receiver makes all the difference in the world.

Chapter 11

BRANCHES, OR YS

It is not uncommon for a trace to branch into two or more segments somewhere along its length. We often call this a Y. If we are dealing with controlled impedance traces we must be very careful how we handle these branches.

Figure 11-15 illustrates three different branch situations. The first (a) illustrates a 50-Ω trace that branches into two individual 50-Ω traces. A designer might incorrectly interpret the design rule that trace impedances must be constant everywhere along the trace as applying here, believing that this approach is correct. In fact, the impedance is not constant everywhere along this trace. Two parallel 50-Ω traces (like two parallel 50-Ω resistors) look like a single 25-Ω trace. Therefore, the 50-Ω trace from the driver "sees" a shift to 25 Ω at the point where the traces Y and there will be a reflection from that point.

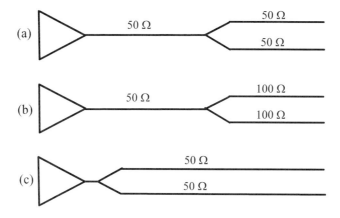

Figure 11-15 Different techniques for handling branches, or Ys, in a trace.

There are two possible ways to handle such a branch correctly. Either one will produce a clean waveform with no reflections. One of these is shown in Figure 11-15b. Design the traces after the Y as 100-Ω traces. Two parallel 100-Ω traces will look like a single 50-Ω trace and there will be no impedance discontinuity or reflection at the Y. The 100-Ω traces, of course, must then be terminated correctly with 100-Ω resistors.

The other way to handle a Y is shown in Figure 11-15c. Bring the branch in very close to the driver so that the impedance discontinuity happens well within the critical length. Then the discontinuity will have little effect.

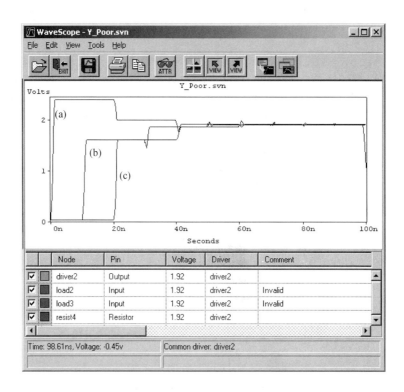

Figure 11-16 Simulation of an improperly constructed branch, or Y. The driver signal is denoted by (a). The reflection from the improperly constructed branch occurs 10 ns later (b). The final signal at the termination is denoted as (c).

Figure 11-16 illustrates the results from a simple simulation of the improper way (Figure 11-15a) to handle a Y. A 50-Ω trace extends 10 ns and then branches into two 50-Ω traces for another 10 ns. The end of each branch is terminated in a 50-Ω resistor. There is an obvious reflection caused by the Y. The pattern is typical of a load (in this case 25 Ω) that is lower that the characteristic impedance of the trace (starting out at 50-Ω, see Figure 11-7). The reflection at the Y (at 10 ns) is evident.

12 CROSSTALK

In Chapter 9 we talked about how current flowing along traces can radiate EMI. In Chapter 10 we talked about how traces couple to adjacent planes, how that relates to return currents, and then to impedance control. In Chapter 16 we will talk about differential traces coupling to each other and the implications that has for signal integrity and impedance control. All of those coupling mechanisms are essentially the same; they rely on the facts that (a) like charges (electrons) repel each other, and (b) flowing electrons can create magnetic fields that induce currents in other conductors. In this chapter we look more closely at the specific coupling mechanism that can occur between adjacent traces on the board itself.

FORWARD VERSUS BACKWARD CROSSTALK

Consider the case where we have two traces adjacent to each other (Figure 12-1). Current (the flow of electrons) flows down one of the traces (we call this trace the *aggressor* or *driven* line). That current will couple into the adjacent trace (called the *victim* line) and create two different noise signals. One of those noise signals will flow in the victim trace in the same (forward) direction as the aggressor current. The other noise signal will flow in the victim trace in the opposite (backward) direction. These two currents have different characteristics and degrees of importance in our circuits.

Consider a current flowing down the driven line (in Figure 12-1) and consider an electron as it passes by the specific point X. Electrons are negatively charged particles. Since like charges repel each other, electrons at that point along the victim line will be "repelled" or will move away from point X. They will move in either direction (arrows labeled (3)), so there will be both a backward and a forward component to this reaction. This kind of coupling is almost exactly what we see in capacitors (electrons flowing onto one plate "repel" electrons from the other plate) so this is often referred to as *capacitively coupled crosstalk.*

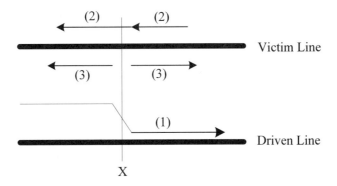

Figure 12-1 There are two causes of crosstalk, inductive and capacitive coupling, which lead to two types of crosstalk, forward and backward.

The current flowing in the direction of the arrow on the driven line will generate a magnetic field around the line. That field will intersect the victim line and induce (or generate) a current going the opposite direction in the victim line. This mechanism is exactly the same mechanism we find in transformers and in motors and generators. In fact, we can think of the driven and victim lines as being the primary and secondary windings of a poorly designed transformer. The induced current (2) always goes in a direction opposite that of the driven current. We call this induced current *inductively coupled crosstalk.*

Note that both these mechanisms (capacitively and inductively coupled crosstalk) depend on the fact that the driven current is changing. No coupling occurs if no (or a constant DC) current is flowing. The faster the current is changing (i.e., the higher the frequency or the faster the rise time) the stronger the coupling between the traces. It also seems intuitive that the closer the traces are, the stronger the coupling will be. Already we have two secrets for reducing crosstalk (coupling): *Slow down the signals and spread the traces further apart.* (Wasn't that easy?)

There is one other point we talk more fully about later. Note that capacitively and inductively coupled crosstalk tend to reinforce themselves in the backward direction. Both create components flowing in the reverse direction. But they tend to cancel each other in the forward direction. Capacitively coupled crosstalk creates a forward component that flows in the same direction as the driven current, but inductively coupled crosstalk flows in the opposite direction. These components almost exactly cancel, particularly in stripline environments. This intuitively leads to another design rule: *If crosstalk is an issue, try to keep all the sensitive traces in a stripline environment.*

Backward Crosstalk There are significant differences in the nature of forward and backward crosstalk. Consider Figure 12-2. This figure is in two rows, Row (1) and Row (2), with four frames each. Look first at Figure 12-2a, Row 1. The two rows of dots at the top represent elements of a victim trace. There are two rows because we are going to let the top one represent backward crosstalk and the other one represent forward crosstalk. The single row underneath them represents elements of the aggressor (driven) trace. There will be a signal, transversing from a low signal level to a high signal level, flowing from left to right along the aggressor trace.

The two rows, 1 and 2, differ in that the victim trace in Row 2 is terminated in a high impedance at the near end. The victim trace in Row 1 extends indefinitely in the backward direction.

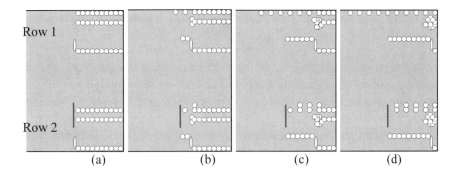

Figure 12-2 Forward crosstalk has a pulse width approximately equal to the rise time of the aggressor pulse with an amplitude that increases over the coupled length. Backward crosstalk has an approximately constant amplitude but increases in pulse width as the coupled length increases.

In the first frame (Figure 12-2a), the aggressor signal is just starting down the trace. In the second frame (Figure 12-2b), the signal has started part way down the trace. In the third frame (Figure 12-2c), the signal has moved well along the trace and by the last frame (Figure 12-2d) the signal has reached the far end of the trace.

Let's look first at backward crosstalk (top pair of dots) for Row 1. Note how, as the driven signal moves down the aggressor trace, each time it passes by a victim element it *kicks* that element to the left. Each element, in turn, starts moving backward before the next element gets kicked. What results is a string of elements moving to the left (backward) stretching out in length. Especially note that by the time the aggressor pulse gets to the far end of the aggressor trace, the backward string of elements is twice as long as the coupled length between the victim and aggressor traces is. This happens because the first element starts moving at the same moment the aggressor trace starts moving. When the aggressor signal reaches the far end, the first element has moved equally far in the opposite direction. But the last element along the victim line doesn't even start to move backward until the aggressor signal reaches the far end. Thus the string of backward-flowing elements stretches twice as far as the length of the coupled region between the traces.

There are two extremely important things to notice here. First, we are talking about *time*, not units of distance. It takes a certain amount of time for the aggressor signal to travel down the trace. We call this the *propagation time* of the signal traveling along the trace. Second, if we assume the signal flows at 6 inches per ns, and each circular element represents .25 ns, then the coupled region between the two traces is 2.5 ns, or 15 inches. The backward crosstalk signal stretches twice as far, or 5 ns.

We can generalize this as follows: *The backward crosstalk pulse is (almost) constant amplitude in magnitude but twice as wide as the propagation time represented by the coupled region.* This backward crosstalk pulse width is what so many people have trouble understanding. The width is not a function of coupling strength. Width is purely a function of the *length* of the coupled region. The statement is worth repeating: *The backward crosstalk pulse width is purely a function of the length of the coupled region.*

Now, Row 2 shows exactly the same thing with one modification. The vertical line at the left of the victim trace represents the beginning of the trace. It might represent a driver or a load. As the first element of the backward crosstalk row gets kicked backward by the aggressor signal, it reflects off the end of the trace and begins flowing in the forward direction. This is not to be confused with forward crosstalk. It is simply a reflected backward crosstalk signal. (One obvious question is this: Could we eliminate this reflection if we correctly terminated this end of the victim trace using the transmission line termination principles we learned about in Chapter 10? The answer is "Yes." More on that later.) In this case, the first element of the backward crosstalk signal reaches the forward end of the trace at the same time the aggressor signal does. But the last element on the victim trace has just started its travel in the reverse direction. Before the show is over, it must travel all the way back to the beginning of the trace, reflect off the beginning of the trace, and travel all the way back to the forward end again—a distance that takes twice as long as it took the aggressor pulse to travel the length of the coupled region. Again, we see that the backward crosstalk pulse width is twice as long as the propagation time down the coupled region.

A word about the magnitude of the backward pulse: If the coupled region is short, the magnitude of the backward signal will be small. It will grow as the coupled length increases. But there is a limit to how high the magnitude will grow until it stops and just levels off. That limit in magnitude is reached when the coupled length is about equal to the so-called *critical length* we have already discussed (Chapter 10). That is, the magnitude of the backward coupled pulse reaches its maximum about when the length of the coupled region equals one-half the rise time of the aggressor signal. This should be considered more as a rule of thumb than as a factual statement, but it is a rule that seems to hold approximately true most of the time.

What that maximum level actually is is a different story altogether. We talk more about that later on.

In summary, backward crosstalk grows fairly quickly to a *constant magnitude* pulse whose pulse width is *twice as wide* as the propagation time down the coupled region.

Forward Crosstalk The second row of dots for each victim trace represents what happens with forward crosstalk. In Figure 12-2b Row 1 or Figure 12-2b Row 2, as the aggressor signal just starts to propagate along the trace, the first element in the forward crosstalk row is kicked forward. It immediately *bumps* into the next element. As the aggressor signal moves forward one more increment, the first two victim elements get kicked further along and bump into the third element.

This continues as the aggressor keeps kicking the forward crosstalk elements along. By the time the aggressor signal reaches the end of its trace, the forward crosstalk elements are all bunched together at the far end of the victim trace.

Note that the forward crosstalk illustration is exactly the same for Row 1 and Row 2. It does not matter whether the victim trace begins at the beginning of the coupled region or extends further back. Since the forward crosstalk signal never travels in the reverse direction, it doesn't care whether there is a reflecting barrier or not.

The quantity of the forward elements represents the magnitude of the forward crosstalk signal. As suggested in Figure 12-2, the forward crosstalk signal will continue to increase in magnitude the longer is the coupled region. Although there is a theoretical limit to how high the forward crosstalk signal can grow, we are never likely to reach that limit on circuit boards (the coupled region can't be long enough). For our purposes, the forward crosstalk signal simply continues to grow larger and larger as the coupled region increases.

There is one other distinctive thing about the forward crosstalk signal. All the elements representing the signal are clustered on top of each other. This represents the width of the forward crosstalk pulse. In theory, the width of the forward crosstalk pulse is no wider than the rise time of the aggressor signal that creates it. The pulse starts to build as the aggressor signal starts to rise, and it has finished building when the aggressor signal has reached its maximum value.

In summary While the backward crosstalk signal is relatively constant in magnitude and has a pulse width twice as wide as the propagation time down the coupled region, *the forward crosstalk signal has an amplitude that continually increases as the coupled region increases and has a fixed pulse width equal to the rise time (or the fall time) of the aggressor signal.* Simulations in Chapter 13 will illustrate these differences quite clearly.

We also show in Chapter 13 that for all practical purposes, forward crosstalk is very small in microstrip environments and virtually nonexistent in stripline environments. Therefore, for the rest of this chapter, we focus only on backward crosstalk.

ESTIMATING CROSSTALK

Calculations We can develop theoretical formulas for calculating crosstalk based on some simplifying assumptions, but the practical reality is that crosstalk is very difficult to calculate, at least without special tools. The HyperLynx tool does a credible job simulating crosstalk, as will be shown in the next chapter. Otherwise, there aren't very many good ways for quantifying what the level of crosstalk might be in your particular design.

There are three things that affect the magnitude of the (backward) crosstalk noise signal:

1. The degree of coupling that exists between the aggressor and victim traces.
2. The distance over which that coupling occurs.
3. The effectiveness of any trace terminations that might exist.

Coupling Referring to Figure 12-3, the degree of coupling is proportional to, and never exceeds the value determined by Equation 12-1:[1]

$$\frac{1}{1+\left(\frac{D}{H}\right)^2} \quad or \quad \frac{H^2}{H^2+D^2} \qquad [12\text{-}1]$$

where D and H are as defined as in the figure. From this, it is apparent that (as we already knew) crosstalk goes down as H goes down and as D increases. So, the first rule in minimizing crosstalk is to create your stackup so that the traces are close to their reference planes, and spread the traces out.

It is instructive to note something else. Trace width does not appear in this relationship. Trace width has only a minor effect on crosstalk. The other two dimensions are far more important than width (at least for reasonable dimensions).

Distance If the coupled region is very short, crosstalk does not have time to develop. Therefore, it stands to reason that the crosstalk amplitude will be small for short coupling regions. We have already discussed the fact that the amplitude is not related to distance for long distances. (Remember, we are talking about backward crosstalk here.) The magnitude builds up as the length of the coupling region in-

---

[1] Johnson, Howard, *High Speed Digital Design, A Handbook of Black Magic*, Prentice Hall, 1993.

creases, up to a maximum amount (reflected in the coupling coefficient formula). At that point the amplitude levels out and stays constant.

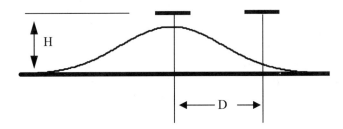

Figure 12-3 Current density under the aggressor trace can couple into an adjacent trace with a coupling coefficient proportional to H^2 and inversely proportional to D^2.

The length of the coupled region that just results in the full amplitude of the crosstalk signal is called the critical length. It is the same critical length we talked about with transmission lines, the length where the round-trip propagation time equals the rise time of the signal. Another way to look at this same question is to look at the round-trip propagation time of the coupled region and test whether that time is greater than the rise time. That is, calculate the round-trip propagation time, T_{RT} (in ns), using the formula discussed in Chapter 2:

For stripline

$$T_{RT} = .085\sqrt{\varepsilon_r} \ \times \ \text{coupled length} \ \times \ 2 \qquad [12\text{-}2]$$

where ε_r = relative dielectric coefficient
and the coupled length is measured in inches.

Then, test if T_{RT} is greater than the rise time, T_r, of the aggressor pulse in nanoseconds (Equation 12-3).

$$T_{RT} > T_r \qquad [12\text{-}3]$$

If $T_{RT} < T_r$, we decrease the coupling ratio in Equation 12-1 by the ratio given in Equation 12-4.

$$T_{RT}/T_r \qquad [12\text{-}4]$$

Otherwise we leave the coupling ratio as it is.

Terminations Finally, we look and see if the victim trace is terminated, and if so how effectively. If two traces are coupled, backward crosstalk will develop. But if the victim trace is a transmission line that is terminated in its characteristic impedance at the near end (Figure 12-4), then the backward crosstalk pulse will be completely absorbed by the terminating resistor and will not reflect forward.

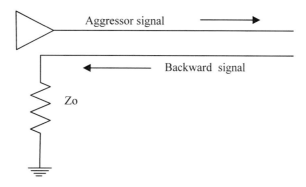

Figure 12-4 Near end termination of the victim trace can absorb backward crosstalk.

In this illustration, there will be no crosstalk at the far end of the victim line because:

- There is minimal or no forward crosstalk anyway.
- The backward crosstalk is absorbed by the terminating resistor (Zo) and will not reflect forward.

This solution may or may not be an effective alternative for your design. First, it may be difficult or impossible to calculate the correct value for a terminating resistor at the near end of the victim line. It may also be impossible to place a terminating resistor there without adversely affecting the operation of the circuit. And, it is frequently the case that individual traces can be both aggressor and victim, depending on when switching occurs. So, although this knowledge may have limited use in eliminating crosstalk, it may be very helpful in making judgments about how destructive crosstalk signals may be to the normal circuit operation.

UltraCAD Calculator UltraCAD has created a freeware calculator for estimating crosstalk based on the preceding discussion (see Appendix H). It is illustrated in Figure 12-5. A detailed Help file explains the operation and the formulas used. UltraCAD does not guarantee any level of accuracy with this calculator, and suggests that it be "calibrated" first against prior designs known to be good and those known to have crosstalk problems. Nevertheless, even though the company does not guarantee any level of accuracy, it is still fun to sometimes compare it against other tools.

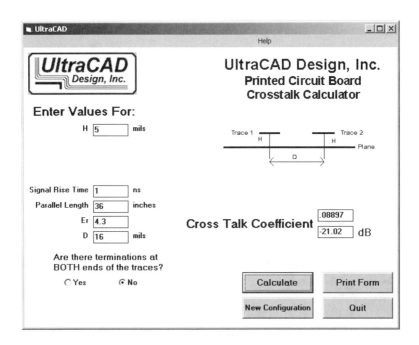

Figure 12-5 UltraCAD's freeware crosstalk calculator.

Figure 12-5 shows the calculated crosstalk coefficient for two 36-inch traces spaced 16 mils apart, each 5 mils above the reference plane, with $\varepsilon_r = 4.3$. The calculated coefficient is .08897. This means that you would multiply the aggressor voltage by this coefficient to calculate the crosstalk voltage on the victim line.

HyperLynx Simulation Tool Figure 12-6 illustrates a HyperLynx model with the same parameters as used with the UltraCAD calculator. (More about how to work with HyperLynx crosstalk models is presented in the next chapter.) The driver voltage in the HyperLynx model is measured as 4.4 volts. Resistor RS(A1) is placed simply to make sure there is no attenuation in the backward crosstalk signal caused by U(A1). Similarly, the load resistor RD(B1) is set to a high value so as not to load down the far-end crosstalk signal. The reverse crosstalk noise is measured as 406 mV at the resistor at the far end (RD(B1)). If we take the 4.4 volts and multiply it by the crosstalk coefficient calculated from the UltraCAD crosstalk calculator (.08897), we get 4.4 x .08897 = 391 mV as the predicted magnitude of the coupled signal. The HyperLynx model predicts 406 mV. Not too bad for a coarse estimating tool.

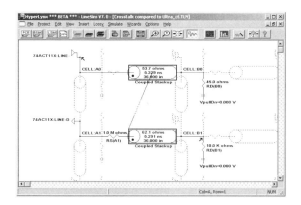

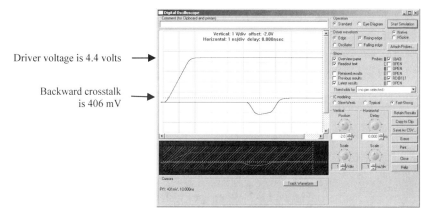

Figure 12-6 HyperLynx model for a pair of 36-inch traces placed 16 mils apart. The model predicts 406 mV backward crosstalk.

DESIGN CONSIDERATIONS

It should be clear by now that the designer has only a few tools to work with in controlling crosstalk. Crosstalk can be reduced by:

1. Using stripline configurations (eliminating forward crosstalk).
2. Using stackups that place traces as close as possible to their reference planes.
3. Separating traces as far apart as possible.
4. Making judicious use of terminations for further reducing crosstalk (but determining the extent is a very complicated process).

Slots in Planes, 3 There is one other crosstalk design consideration that designers need to remember when laying out boards. Figure 12-7 illustrates yet another case where routing traces across a slot in a plane can have serious effects (refer back to Figures 9-7 and 10-18). The signal traces may meet all the criteria specified for good crosstalk control, especially separation. Remember, however, that the signal *returns* want to travel directly underneath the traces. When the returns meet the slot, they must travel around the slot and then come back to their position under the traces. During this path, the returns are very close together. So, while it is true that the traces do not crosstalk with each other, the returns might!

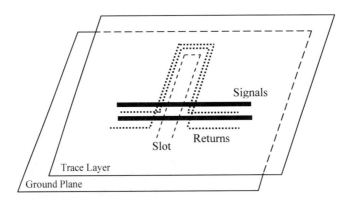

Figure 12-7 When return signals have to divert around a slot or discontinuity in a plane, crosstalk problems can result.

When this board goes to test, there might be a crosstalk problem. The engineers may spend hours going over the board, trying to find that problem. If they are not alert to the existence of the slot in the plane, they may never find the fundamental problem. After several other fixes and a redesign that no longer has the slot, the board may finally meet the desired specifications. All the other fixes may be credited with solving the problem, and a new set of erroneous design rules will result. Don't cause these kinds of problems for your engineers and companies during the design stages of your boards.

Guard Bands Guard bands are sometimes used to isolate traces from each other. The idea is to place a trace, called a *guard band*, between the aggressor and the victim traces. The guard band trace is grounded. Thus, it acts as an electrical shield between the other two traces, much like the shield in a shielded cable might.

The most common application for guard bands is when crosstalk is an issue. The idea is that any coupling that takes place will be to ground (the guard band) and not into the next, more sensitive signal trace.

Although there is not uniform agreement on this issue, most people feel that guard banding is ineffective at best and can actually be harmful at worst. There are several issues that can be of concern here. The first is how and where to ground the guard trace. It certainly must be grounded at one end, but many people say it should be grounded at both ends. The problem with this solution is that now the possibility for ground loops has developed. That is, there is the possibility that stray currents may find their way onto the guard trace at one end and propagate down the guard trace and back to ground at the other end. Ground loops can have very serious effects on performance, not the least of which is the creation of EMI problems.

Still others say that guard traces should be "stitched" to ground every so often along their length. The stitch spacing is an additional point of disagreement among various engineers.

Finally, there are people who say that guard banding actually can make crosstalk worse. The signal from one trace will couple into the guard trace and then that signal will couple from the guard trace to the next signal, making the crosstalk between the two signal traces worse than if there had been no guard band at all.

We know of no definitive studies that clarify this issue one way or the other. The anecdotal evidence that exists supporting guard banding is often explained away by saying that the benefit does not come from the guard trace itself, but from the simple act of spreading apart the two signal traces to make room for the guard band. The benefit comes from separating the traces, not by putting another trace in between them.

We do not have a clear answer here, but we can observe that the two times our customers have insisted on placing significant amounts of copper between signal traces, the boards didn't work. They had serious crosstalk problems. The redesigned boards, without the copper floods, worked significantly better. We don't recommend guard banding to our customers.

13 CROSSTALK SIMULATIONS

In this chapter we look at a series of crosstalk simulations based on the HyperLynx LineSim tool. This tool has an excellent capability for simulating a hypothetical crosstalk situation without the necessity of building up a board design (model) or requiring actual IC models. Therefore, it is a useful tool for looking at and explaining concepts.

BASIC MODEL

Figure 13-1 illustrates a basic coupled model. The HyperLynx model work surface has the various icons already placed on it. You activate them by either selecting or deselecting them. As constructed in Figure 13-1, there are two nets in this basic model. The first consists of a 74ACT11X line driver (U(A0)) connected through a transmission line to a 53-Ω resistor (RD(C0)). The second net consists of a 74ACT11X device set up in receive mode, connected through a transmission line to a 53-Ω resistor (RD(C1)).

There is provision for another transmission line between each IC and the actual transmission lines shown in the model. This provision is shorted out in this model. The reason for that will become evident later. Each coupled transmission line has a 53.3-Ω characteristic impedance and is 67.5 inches long. HyperLynx calculates the delay through each line as approximately 10 ns. The reason for the extraordinary length will also become apparent later.

In this model the transmission lines are based on a hypothetical stackup illustrated in Figure 13-2. The important parameters to note are that this is a microstrip configuration, with 8 mil trace widths separated by 5 mil spacing, with the

235

traces 5 mils above the underlying plane. The relative dielectric coefficient is set at 4.3.

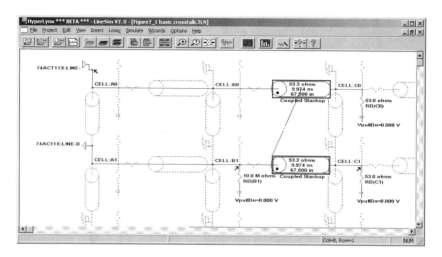

Figure 13-1 A basic HyperLynx model for crosstalk.

You will note that there are three "arrows" in Figure 13-1, pointing to U(A0), RD(B1), and RD(C1). These correspond to oscilloscope probes, or points that will be displayed on the oscilloscope screen during the modeling. The 10.0-MΩ resistor (RD(B1)) is simply placed on the model as a place to put a probe so we can display the signal at that point

Figure 13-3 shows the result of the model. The model is set to use an edge-driven driver waveform with a rising edge and a "Fast-Strong" (fast rise time) step-function signal. The vertical scale is set to 1 volt/division and the horizontal scale to 5 ns/division. Three waveforms, corresponding to the three probe points (arrows), are shown on the screen.

Recall in Chapter 12 we said that forward crosstalk was small in microstrip configurations and virtually nonexistent in stripline configurations, and also that forward crosstalk increases with increasing length. The reason for the extraordinary lengths of these traces is to generate forward crosstalk signals that are clearly visible. Later in this chapter we use more realistic lengths to see what might really happen on normal circuit boards.

Chapter 13

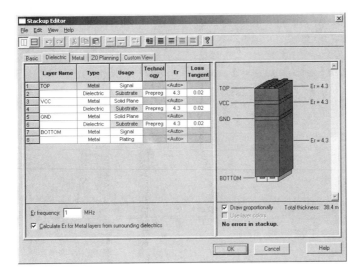

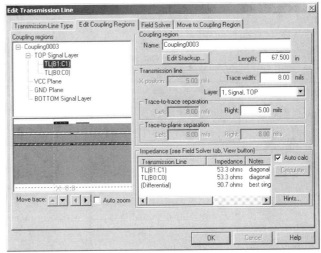

Figure 13-2 Screens for defining and editing the stackup and coupling region between the coupled transmission lines in Figure 13-1.

The driver signal (a) is shown rising immediately on the left axis. There is a coupled backward crosstalk signal (b) that immediately reflects off the receiver and starts forward down the victim trace. Note that this signal is approximately 20 ns wide, just twice as wide as the coupled region (exactly as predicted in Chapter 12).

Ten ns after the aggressor pulse rises (the delay of the coupled region) the forward crosstalk pulse (c) shows up as a negative-going signal at RD(C1). This is a sharp pulse whose width is approximately 2 ns (approximately the rise time of the aggressor pulse, again just as predicted in Chapter 12). Immediately following the forward crosstalk signal is the 20-ns backward crosstalk pulse showing up at RD(C1).

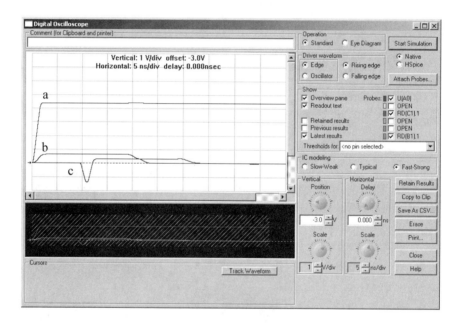

Figure 13-3 Actual results of the model. Traces a, b, and c represent the aggressor signal, the backward crosstalk signal at at RD(B1), and the combined forward and backward signals at RD(C1), respectively. Note how the backward crosstalk pulse width (20 ns) is twice the coupled length (10 ns).

The backward crosstalk signal at RD(C1), the far end of the victim trace, is exactly one-half the amplitude of the backward crosstalk signal at RD(B1), the near end of the victim trace. They are the same signal. But the signal at RD(C1) (far end) is delayed by 10 ns (the length of the line) and is reduced by 50% at RD(C1) because of the voltage divider action formed by the 53-Ω transmission line and the 53-Ω terminating resistor. That is, one-half the voltage at RD(B1), at the near end, appears across RD(C1) and the other half appears across the trace (transmission line) itself.

ADD AN UNCOUPLED REGION

Now let's add a little complexity. Figure 13-4 is the same model with the addition of another 5-ns long uncoupled transmission line section placed in front of the coupled region. Think of this as a pair of traces that are widely separated over part of their length (5 ns), then coupled together over another part of their length (10 ns).

Note how the backward crosstalk signal (b) shows up at RD(A1) 10 ns after the driver signal (a) rises. That 10 ns consists of 5 ns for the driver signal to get to the coupled region, and then another 5 ns for the backward crosstalk signal to travel *back* from the coupled region to RD(A1). This backward crosstalk pulse width is still 20 ns wide, twice the length of the coupled region. It is also the same magnitude as before.

The forward crosstalk signal arrives (c1) at RD(C1) 15 ns after the driver signal rises. That 15 ns is made up of the 5 ns it takes the driver signal to get to the coupled region and then 10 ns for the forward crosstalk signal to travel through the coupled region. The forward crosstalk signal is the same magnitude as before.

It takes about 25 ns for the backward crosstalk signal (c2) to arrive at RD(C1). That 25 ns consists of the 5 ns for the driver signal to get to the coupled region, plus 5 ns for the backward crosstalk pulse to travel backward to U(A1) where it reflects forward again, plus 5 ns back through the uncoupled length of transmission line, plus the final 10 ns to travel through the coupled region. By inserting this 5-ns length of uncoupled transmission line into the model, we have completely separated the forward and backward pulses so that they arrive at the far end on the victim trace at entirely different times. Think, now, of the engineer who might be troubleshooting this simple circuit. The crosstalk signals are showing up at the far end of the victim trace at different times, and the backward crosstalk signal is a different amplitude depending on where you measure it.

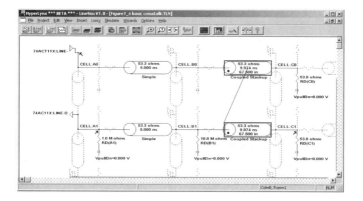

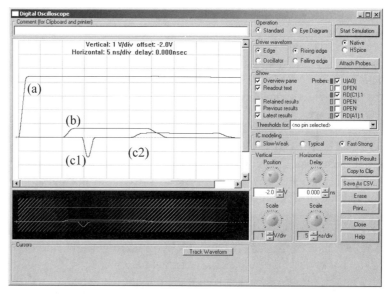

Figure 13-4 We are adding a 5-ns uncoupled region in front of the coupled region. This effectively separates the backward and forward crosstalk signals at the far end of the victim line.

EFFECT OF LENGTH

In Chapter 12 we pointed out that as the coupled region increases:

- The forward crosstalk pulse amplitude increases.
- The forward crosstalk pulse width remains constant.
- The backward crosstalk amplitude remains constant.
- The backward crosstalk pulse width increases with the coupled region.

These effects are easily simulated with the HyperLynx simulator. Refer back to the model shown in Figure 13-1. This time we are going to run the exact same model three times, with coupled region lengths equal to (a) 33.75 in., (b) 67.5 in., and (c) 135 in., respectively. This will correspond to coupled regions of approximately 5 ns, 10 ns, and 20 ns, respectively. Figure 13-5 illustrates the results.

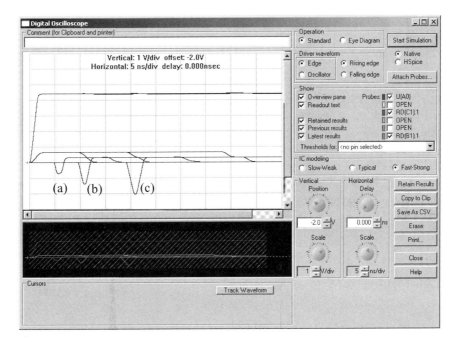

Figure 13-5 Changing the coupled length causes the backward crosstalk pulse to increase in width and the forward crosstalk pulse to increase in amplitude. The modeled lengths are (a) 33.75 in., (b) 67.5 in., and (c) 135 in.

Note how the amplitude of the forward crosstalk pulse increases as the length of the coupled region increases. Note also, however, that the pulse width of the forward crosstalk pulse remains constant, regardless of the coupled length.

Notice how the amplitude of the backward crosstalk pulse does not change as the coupled region changes, but its pulse width does. The backward crosstalk pulse widths are 10 ns, 20 ns, and 40 ns, respectively, twice the length of the coupled region for each case.

Stripline

All the previous models assumed a microstrip configuration. Let's now do similar simulations, this time using a stripline configuration. Figure 13-6 illustrates the setup. We are using effectively the same trace dimensions as before (8 mil wide, 5 mil spacing, 67.5-inch length), but this time the traces are on an inner stripline layer (part of a dual stripline stackup). Because the traces are now in a stripline environment, some of the trace characteristics change. The characteristic impedance is now shown to be 57.3 Ω, and the delay down the line is just under 12 ns.

Figure 13-7 illustrates the result of the simulation. The result looks almost exactly like Figure 13-3, with slightly different pulse widths and voltage levels because the stripline parameters are slightly different. There is one very important exception. There is no forward crosstalk pulse, even with a 67.5-inch (5.6-*foot*) line.

This illustrates the point made earlier that in stripline environments the capacitive and inductive components of forward crosstalk are almost exactly equal and opposite—and they cancel out. Thus, if we are concerned about crosstalk, put all the traces in a stripline environment and one of the crosstalk components goes away. It is as simple as that.

Stripline with Terminations

Now, if you liked that, you will positively love what comes next. In Figure 13-8 we modify the stripline environment a little further and add a terminating resistor, shown as RD(A1), at the near end of the victim line. This terminating resistor will absorb the backward crosstalk flowing toward U(A1), so there will be no reflection from the near end. Now, when we do the simulation (Figure 13-8) we see (a) the aggressor pulse leaving the driver, and (b) the backward crosstalk pulse as it propagates back to RD(A1). However, virtually no pulse is reflected toward the far end of the coupled region (RD(C1)).

Chapter 13

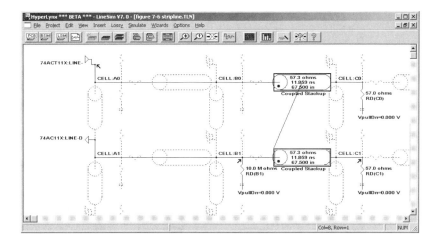

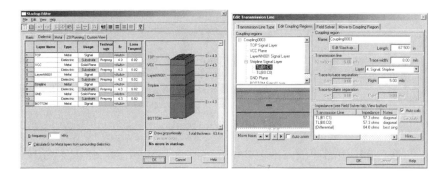

Figure 13-6 HyperLynx model set up for stripline.

Look at what we have here. Traces that are 5.6 *feet* long, spaced 5 mils apart, and there is virtually no crosstalk at the far end, (and you thought crosstalk was a difficult problem).

The practical reality is that we may not be able to place terminating resistors on our boards or in our circuits as easily as we can in this model. If not, we probably can't eliminate crosstalk at the far end completely. And we really haven't eliminated backward crosstalk at the near end, just absorbed it so it won't reflect back. So it is not correct to say that we just eliminated crosstalk, but we certainly have knocked it down some.

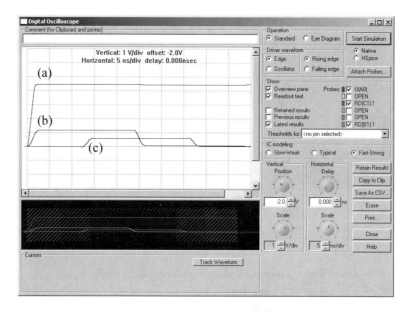

Figure 13-7 Stripline results showing drive signal (a), backward crosstalk signal at RD(B1) (b), and backward crosstalk signal at RD(C1) (c). Note the absence of a forward crosstalk pulse.

It is this final interaction of termination strategies and backward crosstalk reflections that make practical assessments of crosstalk very difficult for circuit design engineers to estimate and plan for.

More Realistic Example

All of these illustrations used unrealistic trace lengths to create noise signals big enough to easily see and analyze, but what do real signals look like? Figure 13-9 shows a simulation of exactly the same setup as Figure 13-6 but with only a 12-inch (2.1-ns) coupled region. The backward crosstalk reflection, when it shows up at the far end (c), is pretty significant, approximately 600 mV. This design, however, is not a very good design for crosstalk control. The traces are 10 mil off the plane and there is only 5 mil spacing between the traces.

Chapter 13   245

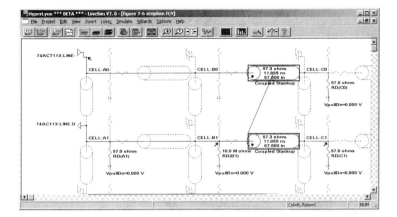

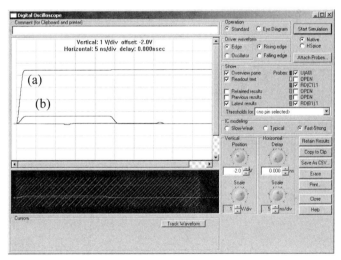

Figure 13-8 Combining a stripline with near-end termination can completely eliminate all crosstalk from the far end of the victim trace.

A better design would have the traces perhaps 5 mil above the planes and then use 16-mil spacing (with our 8-mil traces). Figure 13-10 illustrates this simulation and shows how dramatically the crosstalk is reduced. Crosstalk at the far end is now approximately 100 mV. (Remember, this simulation is using no terminating resistors at the near end.) Whether this level of crosstalk is "good enough" is up to the circuit design engineer to decide.

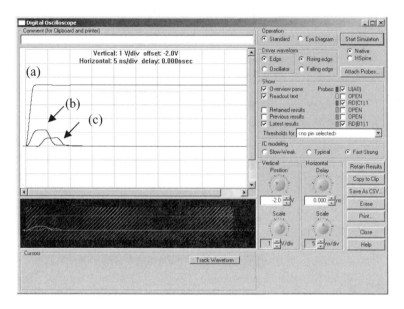

Figure 13-9 Simulation of the stripline example with only a 12 inch coupled region. Backward crosstalk (c) is still somewhat large.

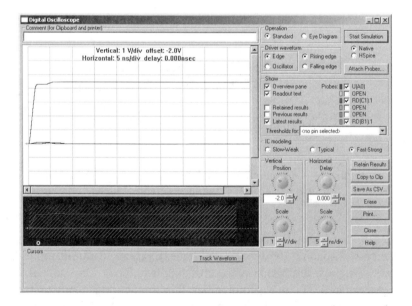

Figure 13-10 Crosstalk model results from a more normal design made with crosstalk control in mind.

SUMMARY

These simulations confirm the steps we need to take to control crosstalk:

- Place traces in stripline environments.
- Minimize the distance between the trace and the plane.
- Maximize the separation between the traces.
- Consider the beneficial effects of near-end terminations.

14 DIFFERENTIAL TRACES AND IMPEDANCE

BACKGROUND

*W*e generally think of signals propagating through our circuits in one of three commonly understood modes: single-ended, differential mode, or common mode.

Single-ended mode is the mode we are most familiar with. It involves a single wire or trace between a driver and a receiver. The signal propagates down the trace and returns through the ground system.[1]

Differential mode (more properly called *odd mode*) involves a pair of traces (wires) between the driver and receiver. We typically say that one trace carries the positive signal and the other carries a negative signal that is both equal to, and the opposite polarity from, the first. Since the signals are equal and opposite, there is no return signal through ground; what travels down one trace comes back on the other.

Common mode (more properly called *even mode*) signals are those that travel in the same direction on both traces. They are generally created by some sort of unwanted noise or variation from unexpected conditions.

Advantages Differential signals have one obvious disadvantage over single-ended signals. They require two traces instead of one, or twice as much board area. However, they also possess several advantages:

---

[1] In truth the signal can return through either or both the ground or power system. I use the singular term *ground* throughout this chapter simply for convenience.

1. Differential circuits can be very helpful in low-signal-level applications. If the signals are very low level, or if the signal/noise ratio is a problem, then differential signals effectively double the signal level:
$$(+v - (-v) = 2v)$$

    Differential signals and differential amplifiers are commonly used at the input stages of very low-signal-level systems.

2. Since differential signals are (by definition) equal and opposite, there is no return signal through any other path. If there is no return signal through ground, then the continuity of the ground path becomes relatively unimportant. If we have, for example, an analog signal going to a digital device through a differential pair, we don't have to worry about crossing power boundaries, plane discontinuities, and so on. Separation of power systems can be made easier with differential devices.

3. Differential receivers tend to be sensitive to the *difference* in the signal levels at their inputs, but they are usually designed to be *insensitive to common-mode* shifts at the inputs. Therefore, differential circuits tend to perform better than single-ended ones in high-noise environments.

4. Switching timing can be more precisely set with differential signals (referenced to each other) than with single-ended signals (referenced to a less precise reference signal subject to noise at some other point on the board). The crossover point for a differential pair is very precisely defined (Figure 14-1). The crossover point of a single-ended signal between a logical one and a logical zero, for example is subject to noise, noise threshold, and threshold detection problems.

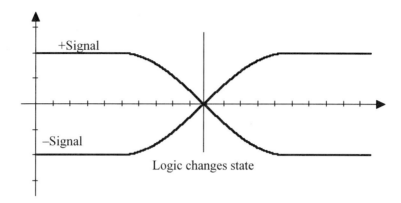

Figure 14-1 Timing with differential signals can be very precise.

Key Assumption There is one very important aspect to differential signals that is frequently overlooked, and sometimes misunderstood, by engineers and designers. Let's start with the two well-known laws that (a) current flows in a closed loop and (b) current is constant everywhere within that loop. Consider the positive trace of a differential pair. Current flows down the trace and must flow in a loop, normally returning through ground. The negative signal on the other trace must also flow in a loop and would also normally return through ground. This is easy to see if we temporarily imagine a differential pair with the signal on one trace held constant. The signal on the other trace would have to return somewhere, and it seems intuitively clear that the return path would be where the single-ended trace return would be (ground). We say that, with a differential pair, there is no return through ground not because it can't happen, but because the returns that do exist are equal and opposite and therefore combine to zero and cancel each other out.

This is a *very* important point. If the return from one signal (+i) is exactly equal to, and the opposite sign from, the other signal (–i), then they combine to zero and there is no current flowing anywhere else (and in particular, through ground). Now assume the signals are not *exactly* equal and opposite. Let one signal be +i1 and the other be –i2 where i1 and i2 are similar, but not exactly equal, in magnitude. The combination of their return currents is (i1 – i2). Since this is not zero, then this incremental current must be returning somewhere else, presumably ground.

So what, you say? Well, let's assume the sending circuit sends a differential pair of signals that are, in fact, exactly equal and opposite. Then we assume they will still be so at the receiving end of the path. But what if the path lengths are different? If one path (of the differential pair) is longer than the other path, then the signals are no longer equal and opposite during their transition phase at the receiver (Figure 14-2). If the signals are no longer equal and opposite during their transition from one state to another, then it is no longer true that there is no return signal through ground. If there is a return signal through ground, then power system integrity *does* become an issue, and EMI may become a problem.

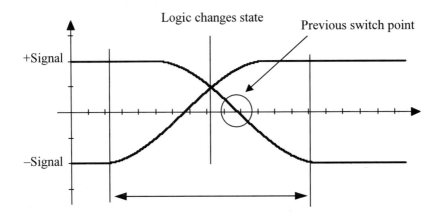

Figure 14-2 The (−) trace is shorter than in Figure 14-1, and it is no longer true that the differential signals are equal and opposite over the range indicated by the arrow. Thus, there will be current flowing through the power system during this time frame.

Let's look at this another way. The square wave in Figure 14-3 is similar to that in Figure 1-13. In a differential signal pair, we might have this signal on one trace, and the opposite signal on another trace. These two signals would then sum to zero (Figure 14-4).

Now consider what happens when we let one trace be slightly longer than the other trace. This is the same thing as having the two signals (the positive and negative signals) being slightly out of phase at the receiving end. Figure 14-5 illustrates the resulting difference signal when this happens. Figure 14-6 illustrates just this difference signal, showing more clearly that it can be very pronounced and also of considerable magnitude for just a very minor difference in phase. The "noise" pulse width in this illustration is equal to the phase shift between the two signals.

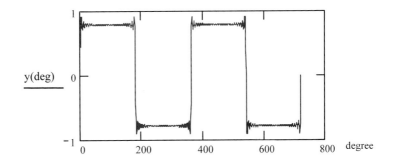

Figure 14-3 Square wave repeated from Figure 1-13.

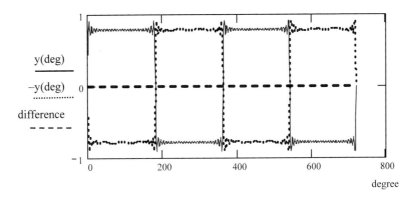

Figure 14-4 The square wave is on one trace and its exact inverse is on the return trace. They combine to zero.

This signal might now be showing up on the ground plane. Not only is it not consistent with our assumption that there are no currents on the ground plane, but the current that now shows up on the plane has sharp rise times, is of considerable magnitude, and can be a serious EMI problem.

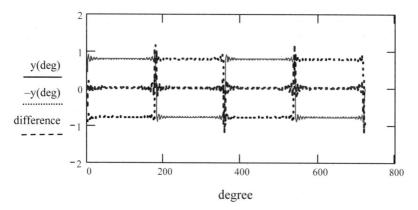

Figure 14-5 If one trace in the differential pair is a slightly different length than the other, a noise signal will be present when they change states.

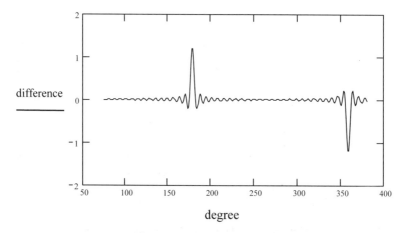

Figure 14-6 A closer look at the noise signal from Figure 14-5.

An interesting question is this: What kind of dimensions are we looking at here before this becomes a problem? A significant part of the answer depends on the rise time of the signal. Even for a poorly defined square wave, a 1- or 2-degree phase shift could be significant.

Assume we have a 50-MHz square wave. That means there are 50×10^6 cycles in one second, or there is a single cycle every 20 ns. If propagation time is 6 inches per nanosecond in FR4, then a one-degree phase shift equates to a 333-mil distance. If we set one degree as our threshold, which might be too much, then the corresponding offsets for selected frequencies would be as follows:

Frequency (MHz)	Offset (mils, or thousandth in.)
50	333
500	33
5 GHz	3

DESIGN RULES

Design Rule 1 This brings us to our first design guideline when dealing with differential signals: *The traces should be of equal length.*

There are some people who argue passionately against this rule. Generally, the basis for their argument involves signal timing. They point out in great detail that many differential circuits can tolerate significant differences in the timing between the two halves of a differential signal pair and still switch reliably. Depending on the logic family used, trace length difference of 500 mil can be tolerated. These people can illustrate their points very convincingly with parts specs and signal timing diagrams. But these people miss the point. The reason differential traces must be of equal length has almost nothing to do with signal timing. It has everything to do with the assumption that differential signals are equal and opposite and what happens when that assumption is violated. What happens is this: Uncontrolled ground currents start flowing that at the very best are benign but at worst can generate serious common-mode EMI problems.

So, if you are depending on the assumption that your differential signals are equal and opposite, and that therefore there is no signal flowing through ground, a necessary consequence of that assumption is that the differential pair signal lengths must be equal.

Common Mode Implications Refer back to the common mode discussion in Chapter 9 and in particular to Figure 9-14. Figure 14-7 is the differential signaling equivalent to the single-ended case shown in Figure 9-14. The currents id and ic represent the signals commonly referred to as differential mode and common mode in differential signaling. In reality, they are correctly called odd mode and even mode, respectively.

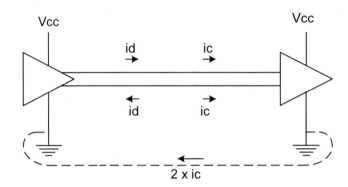

Figure 14-7 Common mode current flows with differential signals.

If the signals are not exactly equal and opposite on the differential traces, we can separate them into their corresponding differential and common (odd and even) mode components as before. Assume again (as we did in Chapter 9) that the plus signal is 10,000 µA (10 mA) and the return signal is 9,950 µA. This is equivalent to the differential (odd) mode component being 9,975 µA and the common (even) mode component being 25 µA. It is easy to see, looking at the problem this way, that if the signal and the return are not exactly equal, there must be a common mode current component that returns through the power system *somewhere*. That current component is uncontrolled, and even though it might be small, it can generate some significant EMI radiations.

Differential Signals and Loop Areas If our differential circuits are dealing with signals that have slow rise times, high-speed design rules are not an issue. Let's assume, however, we are dealing with fast rise time signals. What additional issues then come into play with differential traces?

Consider a design where a differential signal pair is routed across a plane from driver to receiver. Let's also assume that the trace lengths are perfectly equal and the signals are exactly equal and opposite. Therefore, there is no return current path through ground. But there *is* an induced current on the plane, nevertheless.

Any high-speed signal can (and will) induce a coupled signal into an adjacent trace (or plane). The mechanism is exactly the same mechanism as crosstalk. It is caused by electromagnetic coupling, the combined effects of mutual inductive coupling and capacitive coupling. So, just as the return current for a single-ended signal trace tends to travel on the plane directly under the trace, a differential trace will also have an induced current on the plane underneath it.

This is *not* a return current, however. All the return currents have cancelled. So this is purely a coupled noise current on the plane. The question is this: If current must flow in a loop, where is the rest of the current flow? Remember, we have *two* traces, with equal and opposite signals. One trace couples a signal on the plane in one direction, the other trace couples a signal on the plane in the other direction. These two coupled currents on the plane are equal in magnitude (assuming otherwise good design practices). So the currents simply flow in a closed loop underneath the differential traces (Figure 14-8). They look like eddy currents. The loop these coupled currents flow in is defined by (a) the differential traces themselves, and (b) the separation between the traces at each end. The loop "area" is defined by these four boundaries.

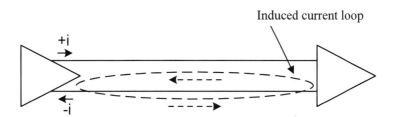

Figure 14-8 Differential traces will couple to a power system plane, even if there are no return currents flowing there.

Design Rule 2 We showed in Chapter 9 that EMI is related to loop area. Therefore, if we want to keep EMI under control, we need to minimize this loop area. And the way we do that brings us to Design Rule 2: *Route differential traces close together*. There are people who argue against this rule, and indeed the rule is not necessary if rise times are slow and EMI is not an issue. However, in high-speed environments, the closer we route the differential traces to each other, the smaller will be their own loop area and the loop area of the induced currents under the traces, and the better control over EMI we will have.

It is worthwhile to note that some engineers ask designers to remove the plane under differential traces. Reducing or eliminating the induced current loops under the traces is one reason for this. Another reason is to prevent any noise that might already be on the plane from coupling into the (presumably) low signal levels on the traces themselves. (I know of no definitive studies that either support or refute this practice.)

There is another reason to route differential traces close together. Differential receivers are designed to be sensitive to the difference between a pair of inputs, but also to be insensitive to a common-mode shift of those inputs. That means if the (+) input shifts even slightly in relation to the (−) input, the receiver will detect it. If the (+) and (−) inputs shift together (in the same direction), however, the receiver is relatively insensitive to this shift. Therefore, if any external noise (e.g., EMI or crosstalk) is coupled equally into the differential traces, the receiver will be insensitive to this (common-mode coupled) noise. The more closely differential traces are routed together, the more equal any coupled noise will be on each trace. Therefore, the better the rejection of the noise in the circuit will be.

Rule 2 Consequence Again assuming a high-speed environment, if differential traces are routed close to each other (to minimize their own loop area and the loop area of the induced currents underneath them), then the traces will couple into each other. If the traces are long enough that termination becomes an issue, this coupling impacts the calculation of the correct termination impedance. Here's why:

Figure 14-9a illustrates a typical, individual trace. It has a characteristic impedance, Z_o, and carries a current, i. The voltage along it, at any point, is (from Ohm's Law) $V = iZ_o$.

Figure 14-9b illustrates a pair of traces. Trace 1 has a characteristic impedance Z_{11}, which corresponds to Z_o, and current i_1. Trace 2 is similarly defined. As we bring Trace 2 closer to Trace 1, current from Trace 2 begins to couple into Trace 1 with a proportionality constant, k. Similarly, Trace 1's current, i_1, begins to couple into Trace 2 with the same proportionality constant. The voltage on each trace, at any point, again from Ohm's Law, now is as shown in Equation 14-1:

$$V_1 = Z_{11} \times i_1 + Z_{11} \times k \times i_2 \qquad [14\text{-}1]$$
$$V_2 = Z_{22} \times i_2 + Z_{22} \times k \times i_1$$

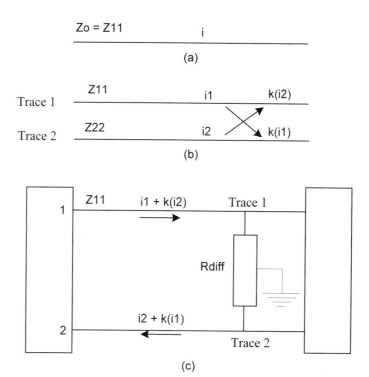

Figure 14-9 Signals on differential traces couple into each other, exactly as crosstalk couples between adjacent traces.

Now let's define $Z12 = k \times Z11$ and $Z21 = k \times Z22$. Then, Equation 14-1 can be written as shown in Equation 14-2:

$$V1 = Z11 \times i1 + Z12 \times i2$$
$$V2 = Z21 \times i1 + Z22 \times i2 \qquad [14\text{-}2]$$

This is the familiar pair of simultaneous equations we often see in texts. The equations can be generalized into an arbitrary number of traces, and they can be expressed in a matrix form that is familiar to many of you.

Figure 14-9c illustrates a differential pair of traces. Recall Equation 14-1:

$$V1 = Z11 \times i1 + Z11 \times k \times i2$$
$$V2 = Z22 \times i2 + Z22 \times k \times i1$$

Now, note that in a carefully designed and balanced situation,

$$Z11 = Z22 = Zo, \text{ and}$$
$$i2 = -i1$$

This leads (with a little manipulation) to Equation 14-3:

$$V1 = Zo \times i1 \times (1-k) \qquad [14\text{-}3]$$
$$V2 = -Zo \times i1 \times (1-k)$$

Note that $V1 = -V2$, which we already knew, of course, since this is a differential pair.

The voltage, V1, is referenced with respect to ground. The effective impedance of Trace 1 (when taken alone this is called the odd-mode impedance of a single trace of a differential pair, or single-ended impedance in general) is voltage divided by current, or, as given in Equation 14-4:

$$Zodd = V1/i1 = Zo \times (1-k) \qquad [14\text{-}4]$$

Since (from above) $Zo = Z11$ and $k = Z12/Z11$, this can be rewritten as Equation 14-5:

$$Zodd = Z11 - Z12 \qquad [14\text{-}5]$$

which is a form also seen in many textbooks.

The proper termination of this trace, to prevent reflections, is with a resistor that has a value of Zodd connected between the trace and ground. Similarly, the odd mode impedance of Trace 2 turns out to be the same (in this special case of a balanced differential pair).

Design Rule 3 The consequence of Design Rule 2 is that the differential pair of traces couple into each other. This coupling impacts the proper termination required if we want to prevent reflections from occurring at the far end of the line. The effect of this is slightly different depending on whether we are looking at the proper differential (odd) mode termination or common (even) mode termination.

Differential Mode Impedance Assume for a moment that we have terminated each trace of a differential pair with a resistor to ground. Since $i1 = -i2$, there would be no current at all through ground. Therefore, there is no real reason to connect the resistors to ground. In fact, some people would argue that you must *not* connect them to ground in order to isolate the differential signal pair from ground noise. The normal connection would be as shown in Figure 14-9c, a single resistor from Trace 1 to Trace 2. The value of this resistor would be the sum of the odd-mode impedance for Trace 1 and Trace 2, or as given by Equation 14-6:

$$Zdiff = 2Zo(1 - k) \text{ or} \qquad [14\text{-}6]$$
$$Zdiff = 2(Z11 - Z12)$$

This is why you often see references to the fact that a differential pair of traces might have a differential impedance of around 80 Ω when each trace, individually, is a 50-Ω trace.

Note this very important consequence: As the traces become closer together, the coupling between them increases. As the coupling increases, Z12 becomes larger, since it is directly related to the coupling coefficient, k. As Z12 gets larger, the differential impedance gets smaller, even though the single-ended impedance (Zo) is not changing. Thus, the more closely two traces are coupled, the lower is the differential impedance. This leads to Design Rule 3: *Differential impedance calculations are necessary with differential signals and traces.*

Common Mode Impedance Just to round out the discussion, common-mode impedance differs only slightly from differential mode. The first difference is that i1 = i2 (without the minus sign). Thus Equation 14-3 becomes Equation 14-7:

$$V1 = Zo \times i1 \times (1 + k) \qquad [14\text{-}7]$$
$$V2 = Zo \times i1 \times (1 + k)$$

and V1 = V2, as expected. The individual trace impedance, therefore, is Zo(1 + k). In a common-mode case, both trace terminating resistors *are* connected to ground, so the current through ground is i1 + i2 and the two resistors appear (to the device) to be in parallel. Therefore, the common-mode impedance is the parallel combination of these resistors, or as given by Equation 14-8:

$$Zcommon = (1/2) \times Zo \times (1 + k), \text{ or} \qquad [14\text{-}8]$$
$$Zcommon = (1/2) \times (Z11 + Z12)$$

Note, therefore, that the common-mode impedance is approximately one-quarter the differential mode impedance for trace pairs.[2]

Design Rule 4 Differential impedance changes with coupling, which changes with trace separation. Since it is always important that the trace impedance remain constant over the entire length, this means that the coupling must remain constant over the entire length. This leads to our fourth rule: T*he separation between the two traces (of the differential pair) must remain constant over the entire length.*

Note that these differential impedance impacts are merely consequences of Design Rule 2. There is nothing really inherent about them at all. The reason we want to route differential traces close together is because of EMI and noise immunity. The fact that this has an impact on the correct termination of "long" traces, and this in turn has an impact on the uniformity of trace separation, is simply a consequence of routing the traces close together for EMI control. (Note: The reason this doesn't happen with other closely routed traces (those subject to crosstalk for example) is that other traces don't have a coupling between them that is perfectly correlated—that is, equal and opposite. If the coupled signals are simply randomly related to each other, the average coupling is zero and there is no impact on the impedance termination.)

DIFFERENTIAL SIMULATIONS

The HyperLynx LineSim simulator can be used to simulate differential waveforms under different conditions. Figure 14-10 shows a typical LineSim setup for a differential signal. The differential driver (U(A0) and U(B0)) at the top of the model drive a pair of traces, each of which has a single-ended impedance (Zo) equal to 50 Ω. The traces are 8 mil wide and are separated by 4 mil, so they are tightly coupled. The traces are each 18 inches (2.5 ns) long.

A differential receiver (U(A1) and U(B1)) is connected at the other end of the traces. The termination of the trace is provided by the resistor RS(A1). The

---

[2]For an interesting discussion about how to terminate both the differential mode and common mode components of a pair of traces, see "Terminating Differential Signals on PCBs," Steve Kaufer and Kellee Crisafalu, *Printed Circuit Design* magazine, March 1999, p. 25.

question is: What value should this resistor be and what are the consequences of picking the wrong value?

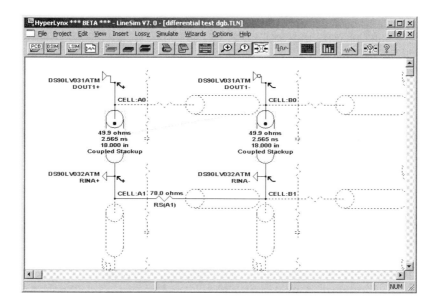

Figure 14-10 HyperLynx LineSim simulation of a stripline differential net.

The HyperLynx termination wizard (part of the LineSim tool) predicts that a termination resistor equal to 78 Ω should be used. Note that this is something less than 2Zo, or 100 Ω. The result of the simulation using 78 Ω is shown in Figure 14-11. The 25-MHz driver signal has a 40 ns period, or is "high" for 20 ns. The signal at the receiver follows that of the driver 2.5 ns later. The signals shown in Figure 14-11 are very clean signals.

Figures 14-12 and 14-13, on the other hand, illustrate what happens when the termination resistor is selected incorrectly. For example, since each line has a single-ended impedance of 50 Ω, a designer may inadvertently select 50 Ω as the terminating differential resistor. Figure 14-12 shows the result. Or someone might simply select a differential termination resistor of 2Zo = 100 Ω. Figure 14-13 illustrates the result of that. In both cases the signal degradations caused by the poorer selection are apparent.

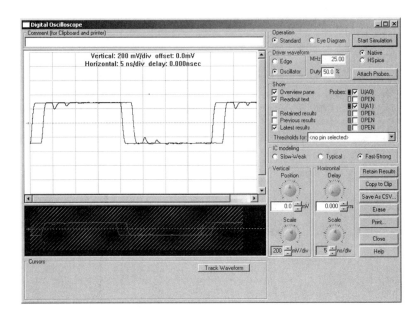

Figure 14-11 Result of differential simulation using 78 Ω.

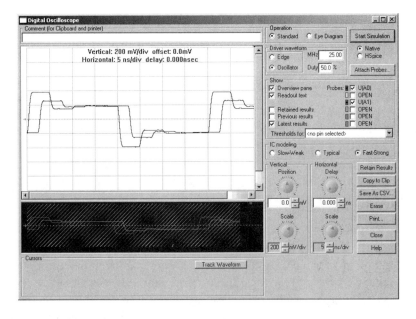

Figure 14-12 Selecting a 50-Ω terminating resistor.

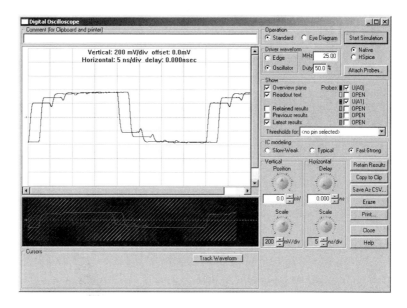

Figure 14-13 Selecting a 100-Ω terminating resistor.

CALCULATING DIFFERENTIAL IMPEDANCE

There are two fundamental types of differential trace configurations: edge coupled and broadside coupled. These are illustrated in Figure 14-14. There does not appear to be a significant advantage or disadvantage to either configuration. The broadside configuration makes it easier to route a pair of traces through a pin field keeping both the length and spacing constant. Broadside coupled configurations, however, typically restrict routing opportunities for other traces. Edge coupled configurations preserve the common X–Y trace routing strategy designers often use with adjacent layers, but then keeping trace lengths equal and traces equally spaced becomes a bigger challenge. Calculations for differential impedance are, of course, different for each configuration.

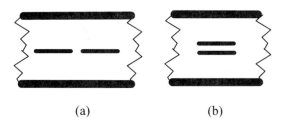

(a)   (b)

Figure 14-14 Edge coupled (a) and broadside coupled (b) differential trace pairs.

Edge Coupled Calculating differential impedance is not easy. In fact, there are only a few tools available that even purport to do this. It is clear from the preceding discussion that any carefully constructed differential signal pair will have a differential impedance something less than 2Zo, where Zo is the single-ended impedance of each of the individual traces. It would not be unreasonable to simply discount 2Zo by 20% and leave it at that. In the simulation shown in Figure 14-10, that would lead to 80 Ω where the HyperLynx wizard calculated an actual value of 78 Ω (pretty close). I am aware of one complex equation for differential impedance that has a stated accuracy level of 20%. (We can *guess* it closer than that.)

National Semiconductor has published some approximate equations for edge coupled differential traces.[3] Higher-end design systems may have a differential impedance calculation ability built into them, so users can make the calculations as they design the board.

Microstrip:

$$Zdiff \cong 2Zo\left(1 - 0.48e^{-.096\frac{s}{h}}\right)$$

Centered stripline:

$$Zdiff \cong 2Zo\left(1 - 0.78e^{-2.9\frac{s}{h}}\right)$$

---

[3] These equations are published in several National Semiconductor publications. See in particular "Transmission Line RAPIDESIGNER Operation and Applications Guide," AN-905, National Semiconductor Corporation.

where:

> Z_o = Single-ended trace impedance
> s = edge-to-edge trace separation
> h = height of the trace above the reference plane (microstrip) or distance between the planes (stripline)

Polar Instruments offers an excellent stand-alone calculator for making both single-ended and differential impedance calculations. It comes in two versions: the SI6000 Quick Solver and an Excel-based spreadsheet plug-in. Figure 14-15 illustrates a Quick Solver solution for a stripline configuration.

Comparison Table 14-1 illustrates the degree of consistency between various calculational tools. For edge coupled differential impedance calculations the degree of agreement between the HyperLynx tool and the Polar calculator is quite remarkable, even recognizing the fact that they both use very advanced field-solving techniques to make these analyses. The National Semiconductor equation agrees within about 9%, which is not bad for simple, approximating equations of this type. Remember, your board house will be lucky to get within 6% or 7% of your differential targets, anyway, for reasons discussed in Chapter 10.

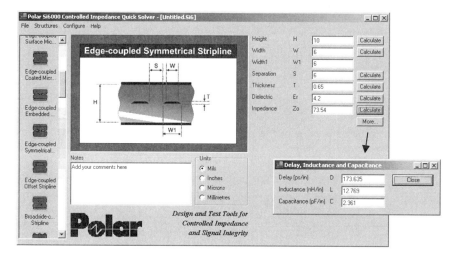

Figure 14-15 Polar Instruments' "QuickSolver" impedance calculator (see at www.polarinstruments.com).

Table 14-1. Comparison of Results

Tool	Zo (Ω)	Zdiff (Ω)
Polar	38.98	73.54
HyperLynx	39.0	73.6
National	39.0*	67.76

* taken as a given

Assumptions:
    h = 10
    w = 6
    s = 6
    t = .65
    ε_r = 4.2

Broadside Coupled Unfortunately, I am not aware of any simple formulas for estimating broadside coupled differential impedance. Higher end tools will do this, including the Polar calculator, but there are no readily available simple tools for doing so.

15 BYPASS CAPS AND DECOUPLING SYSTEMS

*T*here are two views regarding bypass capacitor decoupling strategies today. They appear to be quite different on the surface, and they lead to some differences in results when taken to their logical conclusions. I tend to call them (a) the traditional approach, viewing bypass capacitors almost as little storage batteries (what I sometimes refer to as "little storage buckets of electrons"), and (b) the power system impedance approach, where the focus is on achieving a power distribution system impedance level below some target maximum at all relevant frequencies. We look at both of them here, and I'll highlight the differences in conclusions between them.

TRADITIONAL APPROACH

Figure 15-1 shows a representative logic circuit. If the input goes low (logical zero) Q1 turns on and pulls the base of Q2 low, thereby also forcing the base of Q4 low. Q3 turns on and the output goes high (logical one). Thus, this logic device is an inverting gate. That is, the output goes in the opposite direction from the input.

In a typical application, logic gates are connected together, as shown in Figure 15-2, with the output from one connected to the input of another. Thus, when the input of the first gate goes low, for example, its output will go high, the input of the second gate then goes high, and the output of the second gate goes low.

Logic circuits are designed so that the signal level of their logical states falls within a known and expected range. That is how the second gate knows that it is receiving either a logical one or a logical zero from the first gate. But what if a logical zero value from the first gate was a high enough voltage level that the second gate either could not "decide" what its level was, or it mistook it for a logical one? A logical system error would then result and the system would likely fail.

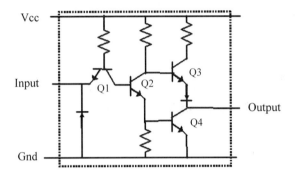

Figure 15-1 Typical logic gate.

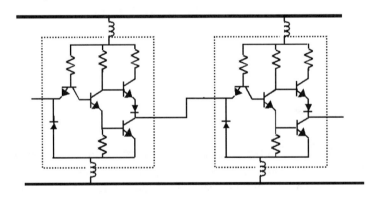

Figure 15-2 Two logic gates connected in series.

Circuits are designed with a wide enough tolerance range so that the devices themselves are unlikely to cause this to happen. But there are external factors that might cause something like this to occur. Consider the situation illustrated in Figure 15-3. Two logic gates are connected in series. There is a common power supply for them, Vcc. There is some level of inductance in the path between the power supply and the devices. (Remember, this was one of the possible contributors to common mode problems identified in Chapter 9.) Several factors contribute to this inductance: the planes, the pads, the vias and escapes connected to the pads, and so on. And remember that current always flows in a closed loop.

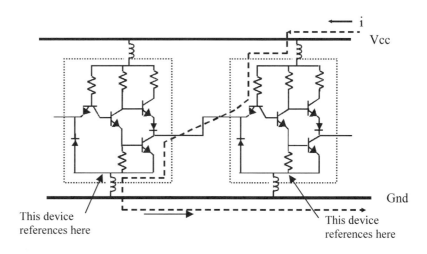

Figure 15-3 Part of current path through first gate when its output goes low.

So when the output of the first device goes low, there is a current path that loops around from Vcc through the power input pin on the second device, through the device to the output of the first device, then through the ground pin, and back to Vcc. In particular, this current flows through the stray inductance shown at the bottom of the device. The transient switching voltage across this inductance is given by the expression $V = L \times di/dt$ (Equation 4-4), where di/dt is the change of current divided by the change in time (where the change in time is the rise time or, in this case, the fall time) of the logical transition. If the fall time is very short, we can have a significant voltage across this inductor even if the inductance is small. Furthermore, if this is a large, several-hundred-pin ball grid array (BGA), there might be a substantial amount of transient current flowing though this inductor.

Now the output signal from this device is the sum of the normal logical zero voltage *plus* the voltage generated across the inductor. If the voltage across the inductor is large enough, this signal might look like a logical one to the next device in the string, and a logic error would occur. Figure 15-4 illustrates how this might happen. As the signal falls from its high to low state, a voltage proportional to the fall time of the signal appears across the inductor. This voltage *adds* to the signal voltage. The result is the dotted curve shown in the figure. The voltage increment across the inductor, which causes the voltage level associated with this device to rise, or "bounce" up to a level higher than it would normally be, is sometimes referred to as *ground bounce*.

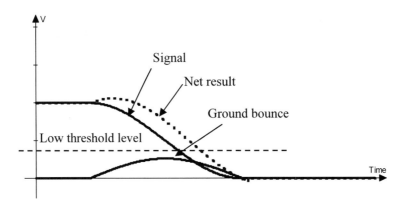

Figure 15-4 Effect of ground bounce on a falling signal transition.

The horizontal line in the figure represents a low threshold. The effect of the ground bounce can be interpreted one (or both) of two ways:

1. At a given point in time after the switch, the signal is higher than it would otherwise be by the amount of the ground bounce.

2. The effective propagation time of the signal slows down; that is, the signal crosses the threshold later than it otherwise would.

Remember that the left side of Figure 15-3 represents only a single gate. There may be a large number of gates all switching at the same time, each adding an increment of current through the stray inductance. If there are a large number of gates, the resulting signal error or delay can prevent the proper operation of the circuit. The issue is this: How do we prevent this from happening?

Figure 15-5 illustrates one way. Place a capacitor around the device that connects between the device and the stray inductances that can contribute to the ground bounce. Figure 15-5 illustrates where. Now, when there is a short-term transient surge of current, the required electrons (remember, current is the flow of electrons) are stored conveniently in the capacitor rather than having to transition from the power supply. This is where the term "little storage buckets of electrons" comes from.

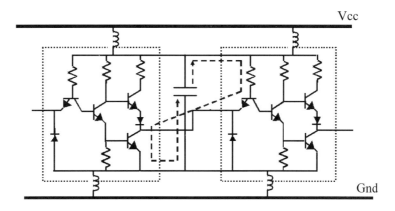

Figure 15-5 A properly placed capacitor stores and then delivers the short-term transient charge needed for switching.

Conceptually, this answer is pretty simple. It does raise a few issues, however. For example, what size capacitor do we use? Do we use more than one capacitor? Where do we place the capacitors? How do we connect them to the circuit board? Figure 15-6 helps us consider some of these questions.

Size and Quantity Looking in from the left side of Figure 15-6, let the first curve represent the required amount of charge needed to complete the logical transition. The area under the curve represents the required charge, and the slope of the curve represents when that charge is required. The furthest curve in (far right) represents what the power system can supply. It can supply a very large amount of charge, but probably not very quickly. The first problem is that power systems usually have a large amount of inductance associated with them, so their response is slowed by that inductance. The second problem is that power systems might be placed some distance away from the need. If electrons can travel at 6 inches in a nanosecond, and if the device switches in a nanosecond, then all the electrons needed during the switching phase need to be located within 6 inches of the device. Half of them need to be within .5 ns (3 inches) away. One-quarter of them need to be within 1.5 inches, and so on.

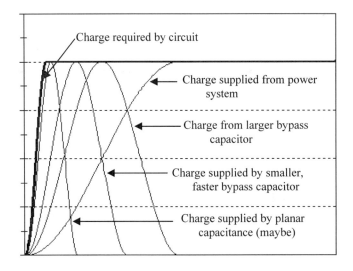

Figure 15-6 Different capacitor types play different roles in supporting transient switching requirements.

The curve next to the power system curve represents what a larger bypass capacitor might be able to provide. It can provide a fair amount of charge, and the inductance associated with it is probably smaller than that of the power system, but the inductance is still not low enough to allow the capacitor to respond quickly enough to meet all the needs of the device.

Adjacent to that curve is another curve for a smaller capacitor. In general, smaller capacitors have less inductance than larger capacitors do, so they can respond faster. But they also store less charge, so they might not be able, by themselves, to handle the requirement.

Finally, planar capacitance, if it exists, might be able to supply charge very quickly, since there is very little inductance associated with a plane. There are, however, several issues associated with planar capacitance:

- It does not exist by accident. It must be designed in.
- The charge is distributed across the plane, not localized.
- The charge on the plane is subject to the same 6-inch rule we just described.

Therefore, planar capacitance, if designed in, can supply charge very quickly, but the amount of charge it can supply is limited.

Finally, it is possible (as shown) that all of these sources of charge together might not be able to meet the requirement. If that should be the case, the device simply cannot switch as fast as it wants to and the system slows down.

The conclusion from all this is that often more than one solution may be required. We may need a larger (say .1 µF) capacitor for bulk charge and a smaller (say .01 or even .001 µF) capacitor for fast response. We may find that we cannot reliably meet the switching needs at all without designing in some planar capacitance. That is why the engineer you are working for may require two (or more) decoupling capacitors for each decoupling requirement on the board. That is why a great many boards now being designed for high-speed applications have planar capacitance designed in (see Chapter 16).

Placement It should be clear from the discussion related to ground bounce that one of the purposes for using bypass decoupling capacitors is to minimize the inductance in the current path associated with the switching device. The way to do this is to place the capacitor as close as possible to the switching requirement. (Indeed, some people would argue that it is not possible to get too close.) The question is this: If there are two capacitors, which one is placed closer to the device?

This answer is straightforward and pretty noncontroversial. Place the smaller, faster one closer to the requirement. It is the one that will be providing the first increment of charge, so it should be placed closest to where the charge is needed.

But since the capacitor is probably smaller than the device, should it be placed closer to the power pin or the ground pin? This question can spark an argument that can go on for days without a conclusion. Engineers can have very strong feelings about this. After listening to all the comments on e-mail forums whenever this topic is raised, my opinion (and why) is as follows:

The objective of the decoupling capacitor is to stabilize the reference voltage (so devices have a stable reference for logical ones and zeros).

- Therefore the capacitor should be placed closest to the reference voltage.
- The reference voltage is ground (zero volts) for both TTL and ECL logic devices, so the capacitor should be placed closest to the ground pin for these logic families.
- The reference voltage for CMOS devices is typically halfway between the lower and higher supply voltages, so the capacitor in these circuits should be placed an equal distance between the power and ground pins on the devices.

Connection Another source of controversy is how to connect the capacitor to the circuit. One school of thought is to connect everything directly to the power and ground planes. Another school of thought is to run traces between the device pins (power and ground) and the capacitor, and then connect the capacitor and the device to the planes with vias at that point (the capacitor). The argument for this second position is that, by definition, there is switching noise between the device and the capacitor, and we want to keep that noise off the planes so it does not interfere with other devices. The argument for the former is that connecting to the planes provides the lowest inductance path, and therefore the fastest switching. Even more recently, some people have argued that vias can have more inductance than traces, and that in fact the traces are the lowest inductance paths.

I know of no definitive studies or results that address this point. After listening to all the arguments and looking at the results of all the boards we have designed, my personal conclusion is that it just doesn't matter very much which way you connect the capacitors to the circuit. What you *do* want to do is minimize the inductance at the pads and vias, however they are connected.

Summary Summarizing the traditional approach:

- Use two or more capacitors per requirement, one for bulk and one for speed.
- Select low-inductance capacitor types.
- Use low-inductance pad and via designs.
- Consider designing in planar capacitance for even faster response.
- Place the faster capacitor closest to the reference voltage.

You might recognize that virtually all application notes from semiconductor suppliers (if they address decoupling at all) suggest variations on this approach. Almost none of them suggest decoupling strategies that are based on the power system impedance approach. This is not particularly surprising since their primary interest is the performance of their particular device. Generally, they have little knowledge about the rest of the system design.

POWER SYSTEM IMPEDANCE APPROACH

Recently, there has been a growing interest in looking at the decoupling issue from a total power system standpoint. The argument goes like this: The traditional approach focuses on the individual IC components and thus is wasteful. What it really achieves by default, and what could be achieved more economically if we addressed it head on, is a (target) low impedance for the power system at all relevant frequencies.

Ideal Response As we place bypass capacitors around the board, by default we are creating a power system impedance curve that is "high" at DC but low at all other frequencies. By focusing on the individual ICs we are not optimizing the system. What if, instead, we focused on the system? Then the needs of the individual ICs would be taken care of by default.

The objective, therefore, is to develop a power system bypass decoupling strategy with the ideal characteristic of infinite impedance at DC and zero (or at least very low) impedance at all other frequencies. This type of response is suggested in Figure 15-7. The ideal, of course, can't be obtained, but the issue is how close we can come to the ideal.

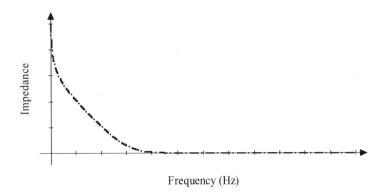

Figure 15-7 Ideal response curve for the power distribution system.

Capacitor Response Let's start with a simple .01 µF capacitor. The impedance, Z, of an ideal capacitor is given by the expression (see Equation 6-1):

$$Z = 1/\omega C = 1/2\pi f C$$
where f = the frequency in Hz

This relationship is shown by the dotted line in Figure 15-8. It is a straight line (on logarithmic scales) and follows the familiar pattern of having very high impedance at very low frequencies and very low impedance at very high frequencies.

Inductance Effects The problem is that a real capacitor doesn't act like this at all. All real capacitors have some inductance associated with them. Some of that inductance is inherent in the structure of the device itself and some is associated with how the device is mounted onto the board. Good SMT devices might have 5 to 10 nH inductance associated with them, so a "real" .01 µF capacitor actually looks more like that shown in Figure 15-9.

The impedance of this capacitor, as a function of frequency, is given by Equation 15-1:

$$Z = \left| 2\pi f L - \frac{1}{2\pi f C} \right| = \left| \frac{(2\pi f)^2 LC - 1}{2\pi f C} \right| \qquad [15\text{-}1]$$

This is the solid curve plotted in Figure 15-8. Note that it follows the curve of an ideal capacitor for lower frequencies, but then as the influence of the inductor comes into play the curve starts to fall more sharply, until it reaches a minimum.

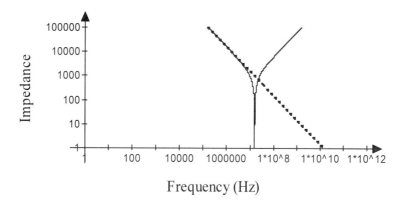

Figure 15-8 Frequency response curve of an ideal and real bypass capacitor.

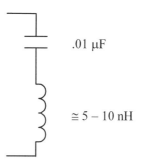

Figure 15-9 Real bypass capacitor.

Then it starts rising rapidly. This minimum occurs when the numerator of the impedance formula goes to zero (refer to Equation 6-13), which happens, in this case (C = .01 µF and L = 10 nH), when:

$$f = \frac{1}{2\pi\sqrt{LC}} = \frac{1}{2\pi\sqrt{(.01)10^{-6}(10)10^{-9}}} = 16 \text{ MHz}$$

We call this minimum the resonance point, or the *self-resonant frequency* of the capacitor. It is the point where the impedance of this LC circuit goes to zero. Not just to a very low value; not just close to zero; it goes to (in the absence of any resistance) absolute zero.

There are two things to recognize in Figure 15-8. The first is the existence of a series resonant point, as just discussed. The second is the behavior of the impedance curve above resonance. We refer to the shape of the curve as *downward sloping to the right* before resonance and *upward sloping to the right* past the resonant point. The significance of these terms is that capacitive circuits are characterized by downward-sloping curves and inductive circuits are characterized by upward-sloping curves. So the impedance curve is inductive past the resonant point. (The formal meaning of that is that the voltage phase shift is positive past the resonant point.)

Some people argue that the effectiveness of a bypass capacitor disappears above resonance; that is, above resonance it acts like an inductor and is therefore useless as a capacitor. This is not necessarily true. Yes, the circuit is inductive above the resonant point, but the overall impedance is still less than that for an ideal capacitor for at least some frequency range above resonance. This is because the capacitor and inductor combined can have (and do have) a lower impedance than the pure capacitor alone until some frequency above resonance where the inductive effect finally and fully takes over.

Multiple Capacitors Now let's start doing something of more interest. Let's look at 200 of these .01 µF capacitors all connected in parallel. This is something like what we do on a circuit board when we add a large number of equal capacitors around the various devices. The combined effect is as if we had a single capacitor with a capacitance of 2.0 µF and a lead inductance of only .05 nH. The response for this set of capacitors is shown in Figure 15-10, along with the curve for a single ideal .01 µF capacitor and a single .01 µF capacitor with associated inductance. Note the set of 200 capacitors still has the same effective resonant frequency. Since each of the individual capacitors has the same self-resonant frequency, the set must have the same self-resonant frequency. (Actually the set has 200 individual self-resonant frequencies, all at the same point.)

Now there are two things to note about Figure 15-10. The first is that at every frequency the impedance of the 200 capacitors is significantly lower than that of an individual capacitor. And, there is a significant range over which that impedance is relatively low. Remembering that our objective is to develop a strategy for achieving a low impedance across a wide frequency range, this curve already suggests part of the approach to getting there—add more capacitors.

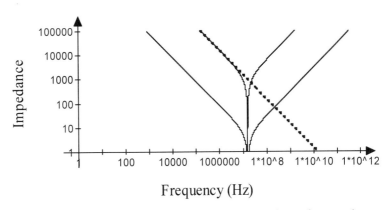

Figure 15-10 Response when we add multiple values of a capacitor.

The second is that there is a significant frequency range (from about 16 MHz to about 200 MHz) where the new impedance curve is inductive, yet the impedance is still well below that for the ideal individual capacitor. Just because the impedance curve is inductive doesn't mean that it isn't meeting the objective of providing low impedance for the power system.

Additional Capacitor Value We can carry our strategy a step further and add additional capacitors of a different value. For example, lets add 20 each 50 µF capacitors. This will provide low impedance at the lower frequencies. These larger capacitors, however, are likely to have larger associated inductance with them. Assume that inductance to be 20 nH per capacitor. The combined effect of these capacitors is shown on the curve in Figure 15-11, along with the effect of the 200 each .01 µF capacitors we have already looked at. We are now beginning to develop a frequency response curve that really begins to look like our objective. If we combined the best parts of the two individual curves, the impedance level would be less than 1 Ω from 100 Hz to approximately 3 GHz.

Planar Capacitance We could, if we wanted to, incorporate into the board stackup some planar capacitance. We do that by placing a power and ground plane adjacent to each other very closely. The smaller the distance between the two planes, the greater the capacitance. A significant advantage to planar capacitance is that there is very little inductance associated with it. We can often assume that it is a pure capacitor over the range of interest.

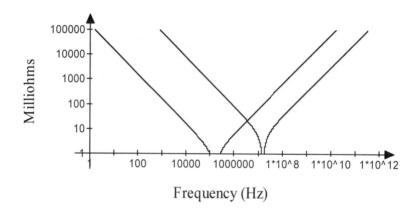

Figure 15-11 Frequency response of two different sets of capacitor values.

A reasonably good formula for approximating planar capacitance is given by Equation 3-10:

$$C = \frac{.2248 \times \varepsilon_r \times A \times (n-1)}{d}$$

where   C is the capacitance in pF
   ε_r is the relative dielectric coefficient
     $\varepsilon_r = 1$ for air
     $\varepsilon_r \approx 4$ for FR4

A = area of the overlapping planes, in^2
d = distance between planes, in inches
n = number of plates

In using this formula, be sure to do the following:

- Be careful to observe the units you are using.
- Account for only the overlapping area of the planes.
- Adjust for such things as holes and voids.

A reasonable number for the amount of capacitance that can be achieved with planar capacitance is in the range of 100 to 300 pF per square inch.

Continuing on with our illustration, lets add 950 pF of planar capacitance to the other capacitors we have assumed. Figure 15-12a shows the result. It would be tempting to look at only the best part of each curve in Figure 15-12a to get something that looks like Figure 15-12b. In this figure, the overall impedance is less than 200 mΩ at all frequencies above 1 kHz. If our objective was to define a set of capacitors that provided a power system frequency response curve less than 200 mΩ at all relevant frequencies, this set would appear to meet the target.

Appearances can be deceiving, however. In Figure 15-12a, look at each point where a downward-sloping line intersects an upward-sloping line. The downward-sloping line is capacitive. The upward-sloping line is inductive. At the point where they cross, their impedances (really their reactances) are equal in magnitude but opposite in sign. This is characteristic of the point where a parallel LC circuit goes into resonance. And at that parallel resonance point, the impedance rises infinitely high. So there is really an infinite impedance peak (pole) at the points where each pair of upward-sloping and downward-sloping curves cross.

The actual frequency response curve is shown in Figure 15-13. The formulas and calculations for this type of problem are quite complex. We at UltraCAD wrote our own calculator specifically to work with this type of problem and to generate graphical results for analysis. The result from that calculator is shown in Figure 15-13. Note, in particular, that the axes are scaled slightly differently between Figures 15-12 and 15-13, accounting for part of the visual difference between them, even though the same circuit is being graphed. Note also that the graphs should extend infinitely high at their peaks and infinitely low at their minimums. They don't because of the graphical limitations associated with the program, not the algebra.

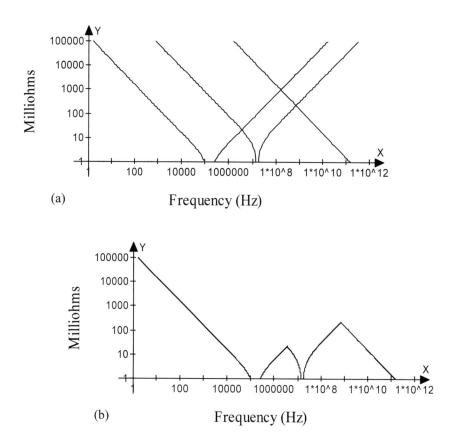

Figure 15-12 Power system frequency curve developed from 20 each 50 µF caps, 200 each .01 µF caps, and 950 pF in planar capacitance.

The implications of this type of curve can be very serious. Most stray frequencies that happened to find their way onto the power system would be shunted to power or ground through an impedance represented by the curve, less than .2 Ω at most frequencies. But any stray noise frequencies (harmonics) that just happened to be near the peaks of the curve shown in Figure 15-13 would *not* be shunted to the power system. They would be free to wander over the system with no attenuation, and would ultimately radiate out off the board. If a board whose power system impedance curve looked like Figure 15-13 were tested for EMI radiation, it would likely fail. We would expect the strongest radiation frequencies to be near the peaks of the curve.

That is why we don't want any peaks in our impedance curves.

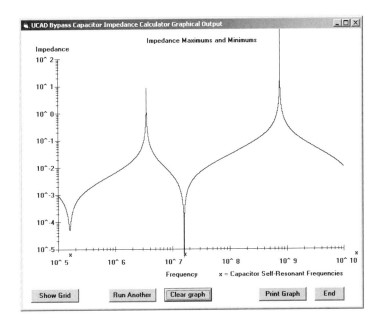

Figure 15-13 Actual response curve of the selected capacitors.

Equivalent Series Resistance (ESR) Well, we know the real world can't be as bad as that shown in Figure 15-13, and it's not. The reason is that bypass capacitors are not pure LC circuits. The equivalent circuit for a typical capacitor actually looks like Figure 15-14. There is a resistance in series with the capacitor and the associated stray inductance. We call this resistance ESR. It is usually very low and is often ignored in most circuit applications.[1]

It turns out ESR has an important role to play in bypass capacitor applications. Most engineers understand that it effectively lowers the peaks and raises the

---

[1] The equivalent circuit for a capacitor could also include a parallel "leakage" resistance around the capacitor. This equivalent parallel resistance is very high and is ignored in most circuits.

troughs in the power system impedance curve. The peak of the impedance curve is inversely proportional to the ESR value; that is, $Z \approx 1/\text{ESR}$. Figure 15-15 illustrates the effect ESR can have on the power system impedance curve shown in Figure 15-13.

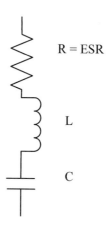

Figure 15-14 A real bypass capacitor includes both inductance and series resistance.

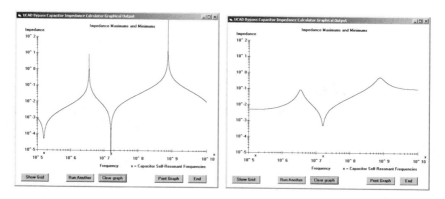

Figure 15-15 Lower ESR (left) will increase the peaks and lower the troughs of the impedance curve; higher ESR (right) will tend to smooth the response.

What a great many engineers do not understand, however, is the impact that ESR has on the overall system impedance curve when a significant number of bypass capacitors are connected in parallel. For example, here are three myths that many engineers firmly believe but are simply not true when ESR begins to influence a circuit.

- *Myth 1:* The system series resonant points (the minimum values of the curve, or the troughs) are at the self-resonant points of the individual capacitors. Truth: When two (or more) capacitors are used with individual series resonant points close together, the system series resonant points are *between* the individual capacitor self-resonant points.

- *Myth 2:* The minimum impedance of the system equals ESR. Truth: The minimum value of the impedance function is *less than* ESR, and it gets smaller as the self-resonant frequencies of the capacitors used in the system get closer together.

- *Myth 3:* Using lower ESR capacitors leads to a better system impedance response curve than using higher ESR capacitors does. Truth: Moderate ESR is better than lower ESR from a system impedance standpoint.

The following analysis demonstrates these points.

Self-Resonant Frequencies Assume a simple capacitor with capacitance C, inductance L, and ESR equal to R. The inductance should be considered from the practical sense—not only the inherent inductance associated with the capacitor physical structure itself, but also the PCB pads and attachment process, and so on. The impedance through this capacitor is given by Equation 15-2:

$$Z = R + j\,\omega L + 1/j\,\omega C, \text{ or}$$
$$Z = R + j(\omega L - 1/\omega C) \qquad [15\text{-}2]$$

where ω is the angular frequency, $\omega = 2\pi f$.

Resonance occurs, by definition, when the j term is zero (see Equation 6-13) as shown in Equation 15-3:

$$\omega L = 1/\omega C \qquad [15\text{-}3]$$
$$\omega^2 = 1/LC$$
$$\omega = 1/\sqrt{LC}$$

The impedance through the capacitor at resonance is R.

Effects of Multiple Capacitors Assume we have n identical parallel capacitors, as earlier. The equivalent circuit of the n identical capacitors is the single capacitor whose values are

$$C = nC$$
$$L = L/n$$
$$R = R/n$$

The impedance of this system is now given by Equation 15-4:

$$Z = R/n + j(\omega L/n - 1/\omega nC) \qquad [15\text{-}4]$$

The resonant frequency of this system is, again, where the j term goes to zero, or where

$$\omega L/n = 1/\omega nC$$

which results in exactly the same self-resonant frequency as before. Paralleling capacitors does not change the self-resonant frequency, but it effectively increases the capacitance, reduces the inductance, and reduces the ESR compared to a single capacitor. The resulting impedance response curve tends to "flatten out" or broaden compared to a single capacitor, as shown in Figure 15-10.

Historically, on circuit boards, circuit designers have used a large number of bypass capacitors of "the same" value. The advantage of this process has been the increased C and the reduced L and R that result. But the capacitor values are typically not exactly equal. Each one has a tolerance.

Parallel Capacitors Take the case of two parallel capacitors of different value, shown in Figure 15-16. Let R1 = R2 = R to simplify the arithmetic. (This assumption does little harm and greatly helps the intuition.) Let us also assume that:

$$C1 > C2$$
$$L1 > L2$$

which means that fr1 (the self-resonant frequency of C1) is lower than fr2.

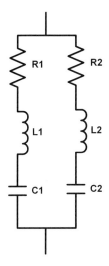

Figure 15-16 Two capacitors in parallel

Now, the combined impedance through this parallel pair of capacitors is given by Equation 15-5:

$$X1 = \omega L1 - 1/\omega C1$$
$$X2 = \omega L2 - 1/\omega C2$$
$$Z1 = R + jX1$$
$$Z2 = R + jX2$$

$$Z = \frac{1}{\frac{1}{Z1} + \frac{1}{Z2}} = \frac{(R+jX1)(R+jX2)}{2R + j(X1 + X2)} \qquad [15\text{-}5]$$

From this, we can derive the real and imaginary terms of the impedance expression (multiply numerator and denominator by $(2R - j(X1+ X2))$ as shown in Equations 15-6 and 15-7:

$$\text{Re}(Z) = \frac{R\left[2(R^2 - X1X2) + (X1 + X2)^2\right]}{4R^2 + (X1 + X2)^2} \qquad [15\text{-}6]$$

$$\text{Im}(Z) = \frac{(X1 + X2)(R^2 + X1X2)}{4R^2 + (X1 + X2)^2} \quad [15\text{-}7]$$

Further, we derive that the magnitude and phase of the impedance term are given by Equations 15-8 and 15-9:

$$|Z| = \sqrt{\text{Re}(Z)^2 + \text{Im}(Z)^2} \quad [15\text{-}8]$$

$$\theta = \tan^{-1}\left(\frac{\text{Im}(Z)}{\text{Re}(Z)}\right) \quad [15\text{-}9]$$

The curve of impedance as a function of frequency is shown in Figure 15-17. It is instructive to look at this curve and the real and imaginary terms of the impedance expression formula together.

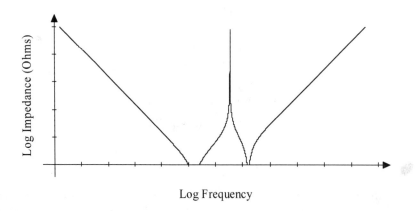

Figure 15-17 Two capacitors in parallel.

Resonance Resonance occurs (by definition) when the imaginary term (Equation 15-7) is zero. This is also the point at which the phase angle is zero. The impedance at that point is simply the real part of the impedance expression.

The imaginary term for Z goes to zero under the two conditions given by Equations 15-10 and 15-11:

$$X1 = -X2 \qquad [15\text{-}10]$$
$$R^2 = -X1X2 \qquad [15\text{-}11]$$

The first condition represents the "pole" between the self-resonant frequencies of the two capacitors. X1 equals –X2 when the reactance term of C1 is inductive (+) and increasing, the reactance term for C2 is capacitive (–) and decreasing, and where the two reactance terms are equal. This is the *anti-resonance* point that occurs at a frequency between fr1 and fr2.

Assuming R is small, the second condition can only occur where either X1 or X2 is small. X1 is small near fr1 and X2 is small near fr2. X1 and X2 must be of opposite sign, since R^2 must be positive. Therefore, these system resonant points must be *between* fr1 and fr2, and they must not be equal to fr1 or fr2 (unless, in the limit, R = 0).

Conclusion The system resonant frequencies are not necessarily the same as the capacitor self-resonant frequencies unless ESR is zero (so much for Myth 1).

It can further be shown that at this point, where the imaginary term is zero and $R^2 = -X1X2$, the real term, and thus the impedance itself, simply reduces to R.

Impedance at fr1 At fr1, the self-resonant frequency of C1, X1 = 0. It can be shown that the phase of the impedance function at this point is

$$\Theta = \tan^{-1}[RX2/(2R^2 + X2^2)]$$

If X1 = 0, then X2 must be negative (capacitive, under the conditions we have been assuming) so

$$\Theta < 0$$

Only in the limit where R = 0 does Θ go to zero.

The *magnitude* of impedance at the point where X1 = 0 is shown by Equation 15-12:

$$|Z| = R\sqrt{\frac{R^2 + X2^2}{4R^2 + X2^2}} \qquad [15\text{-}12]$$

This is *less than* R for any positive (i.e., R > 0) value of R. In the limit, it is equal to R for R << X2 and equal to R/2 if R >> X2.

The results are exactly symmetrical if we are looking at fr2, the point where X2 = 0.

Conclusion *The minimum value for the impedance function is at a frequency other than the system self-resonant frequency and also at a frequency other than the self-resonant frequency of the capacitor. It is less than ESR (R) when two capacitors are connected in parallel. Further, the minimum value declines as X2 gets smaller, or as the self resonant frequencies of the capacitors are moved closer together, or as the number of capacitors increases.*

Impedance at Anti-resonance If we let X1 = –X2, then Im(Z) goes to zero, by definition. This is the anti-resonant point between fr1 and fr2. At this point, the system impedance (Z) is given by Equation 15-13:

$$Z = \frac{R}{2} + \frac{X^2}{2R} \qquad [15\text{-}13]$$

For small values of R, this is inversely proportional to R and can be a very large number if R << X. This is why there is concern about very high impedances at the anti-resonant point. If R, on the other hand, is only in the range of .1 or .01, then this number might be more manageable. But note that if R >> X, then at the anti-resonant point Z ≅ R/2. This is less than the impedance at the system resonant point. Do not overlook the significance of this result. The impedance at the anti-resonant *peak* can be *less* than the impedance at the system resonant point!

So consider this: If Z equals (approximately) R at the minimum (the system resonant point), under what conditions is Z also equal to R at the maximum (the system anti-resonant point)? Remember R equals the capacitor's ESR. Under those conditions, the impedance curve will be (at least approximately) flat. It turns out that Z at the anti-resonant frequency exactly equals R if:

R = X1 = –X2

Therefore, we can achieve a (relatively) flat impedance response curve if we position our capacitor values such that, at the anti-resonant points, X1 = –X2 = ESR.

This has a very significant consequence. As ESR gets smaller, then, for a flat impedance response, X1 and X2 must be smaller at the anti-resonant points. This means that fr1 and fr2 must be closer together. And *this* means that as ESR gets smaller, it requires more capacitors to achieve a relatively flat impedance response.

Conclusion Low values of ESR require more capacitors to achieve a smooth response. Moderate values of ESR lead to smoother response curves with more manageable anti-resonant peaks (so much for Myth 3).

General Case Analysis As we add more values for C, the algebra associated with these kinds of analyses gets very difficult. We at UltraCAD wrote our own program so we could look at various capacitor configurations and see what happens in a more real-world situation (see Appendix H). The following examples illustrate some of the results we achieved using that calculator.[2]

Figure 15-18 shows the impedance curves for three different sets of bypass capacitors used to design a power system. The sets included 100, 150, or 200 individual bypass capacitors, respectively. The capacitor values were carefully selected so that their self-resonant frequencies were spread evenly over a defined frequency range (evident in the figure). The graphical output covers a frequency range from about 1.0 MHz to about 1,000 MHz. To this extent, the example is contrived, but it illustrates what could be achieved under the most controlled of conditions. It is representative of what would happen over more general conditions. All capacitors had the same ESR, .01 Ω.

What is apparent from the figure is that the impedance curve improves as the number of capacitors increases. More important, each curve is *everywhere* better than the curve above it. That is, the peaks of the 150-capacitor curve are lower than the troughs of the 100-capacitor curve, even though the ESR is the same in both cases. This alone would seem to disprove Myth 2.

---

[2]See also Brooks, "ESR and Bypass Capacitor Self-Resonant Behavior: How to Select Bypass Caps," available at www.ultracad.com.

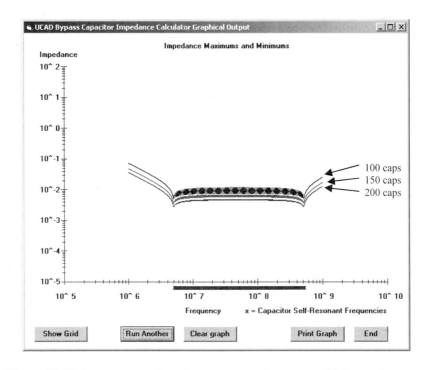

Figure 15-18 Power system impedance response improves with increasing numbers of capacitors.

Figure 15-19 shows the same 150-capacitor curve as Figure 15-18, with ESR equal to .01 Ω, along with a second curve for the same 150 capacitors with ESR of .001 Ω. Although the .001-Ω ESR curve has lower troughs than the other curve, it also has higher peaks. One would be hard-pressed to argue that the lower ESR curve was the better power system impedance response curve. So much for Myth 3.

Consequences So what does all this show us? First, it shows that it is possible to design a power system impedance response curve for any system that can meet an arbitrary target impedance at all frequencies. This type of analysis also suggests that this can probably be done with a smaller number of capacitors, overall, than we would use using the traditional approach to bypass capacitor decoupling. A further advantage of this approach is that, at least conceptually, we could calculate the exact shape and value of the response curve at all relevant frequencies. Obviously, it is not the board designer who would design this kind of decoupling strategy; it is the circuit and system design engineer. But this design approach would lead to slightly different conclusions than would the traditional approach.

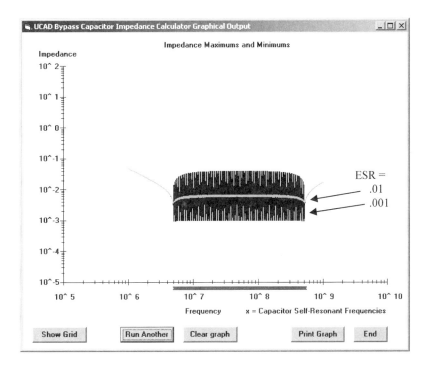

Figure 15-19 Power system impedance response improves with more moderate ESR.

Those conclusions would include the following:

- You must pick the capacitor type carefully, both from the standpoint of capacitance value and of self-resonant frequency.
- Intentionally designed-in planar capacitance is almost a necessity.
- It is best to have a wide range of values and self-resonant frequencies (taking into consideration both the inductance associated with the capacitors and that associated with the pads and the assem-

bly processes mounting them on the boards), rather than to have smaller clusters of identical values.
- Moderate ESR is better than low ESR.
- Capacitor placement is not particularly important, except that the electrons still must get (traveling at 6 in/ns) to where they are needed in a timely manner.

SUMMARY

The jury is still out on the relative merits of these two approaches to bypass capacitor decoupling. Personally, I believe they both have a lot to offer. Overall, I like the logic behind the power system impedance approach, and the results of the quantitative analyses are hard to ignore. Intuitively, however, it is hard to ignore that (especially large) logic devices require a lot of charge support at the moment of switching. The traditional approach argues that you can't place charge supporting capacitors too close to the requirement, something I also believe in.

I believe the very best approach to capacitive decoupling is a combination of these two approaches. Fundamentally, use the power system impedance approach, but then make sure there is enough decoupling at every pin to meet each device's requirement.

16 POWER SYSTEMS

Designers are sometimes confused about how to deal with multiple power planes on printed circuit boards and how to deal with their separation. We often talk about the horizontal separation between planes as a *split,* and there are some design rules that may apply when designing in the area of splits.

Before we talk about how to deal with splits between planes, it is instructive to talk about why we have separate power planes in the first place—to accommodate separate power supplies. One reason we might have separate power supplies is because we might have different voltages in our circuits. The issue of separate voltages might thus be a reasonable place to start.

POWER SUPPLY VOLTAGES

On any given board, or in any given application, we may find DC power supply voltages of 5 V, 3.3 V, 2.5 V, 1.8 V, and even 1.2 V. Older designs have included 12-V supplies, and certain types of circuits may have special design requirements up to thousands of volts.

There are a variety reasons these different voltages exist. One reason relates simply to the semiconductor technology used in the fabrication of the IC making up the circuit. Newer technologies and newer physical semiconductor structures often have inherently lower saturation and switching voltages, leading to lower overall supply voltage requirements for their operation.

A primary driving force for lower voltage relates to the power required for the circuit's operation. If the same circuit function can be achieved with a lower voltage supply, a power saving is usually achieved. Power is the product of voltage times current. Therefore, the power reduction can be directly proportional to the voltage reduction. However, if the voltage reduction also leads to a lower current requirement (e.g., through Ohm's Law), the power reduction can be proportional to the square of the voltage reduction. This has obvious implications for battery-operated circuits and the thermal management of larger, more complex circuits.

Another driving force for lower operating voltages is rise time. If, for example, a technology can switch (dV/dt) a 5-V signal in 2 ns, perhaps it can be modified to switch a 2.5-V signal in half that time, with no other technological changes required.

Finally, different circuit functions may require special voltages. These functions might include such things as transmitter stages, high voltages required by cathode ray tubes and laser printers, switching requirements for electromagnetic relays and machine tools, and audio speaker requirements.

Different voltages require separate regulation circuits. This factor would define a minimum number of regulated power supplies, one for each individual voltage. However, there may be additional, multiple regulated supply sections for any given voltage. For example, we may have different regulated supplies for the analog and digital sections of a board, even if they both have the same overall power supply voltage requirement.

We may also have different regulated supplies for different or sequential stages of the same circuit. For example, if there are different circuit paths for the individual R, G, and B components of a video signal, each path might have its own regulated power supply, even though the individual power requirements are identical. In extreme cases, some engineers design regulated supplies for each stage of a signal path through a circuit.

The reason for multiple regulated supplies is usually noise isolation. All signals flow in a closed loop. For every signal flowing down a trace, there is a return signal coming back. The return signal is usually on the plane closest to the trace, and (at least in fast rise-time circuits) is usually as close as possible to the trace.

Although return current density is greatest underneath the trace, current gradients spread out on the plane beyond the edges of the trace (see Figure 16-1).

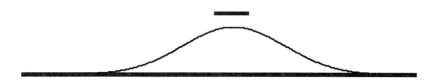

Figure 16-1 Current distribution of a return signal spreads out on the plane underneath the signal trace.

Here is a point that is important to understand: Current is the flow of electrons. Figure 16-1 shows the variation in current density near a trace. It necessarily, therefore, illustrates the *electron density* in the same region. Electrons have charge. Charge density is voltage. Therefore, there is also a *voltage density* (actually a voltage gradient) that occurs near a trace.

Variations in voltage gradients can occur for several reasons. One is high-frequency return current, as suggested in Figure 16-1. Another is simply the flow across the plane of current from the regulated power supply required for charging (and discharging) the various bypass (and planar) capacitances associated with the circuit. Voltage gradients across planes of as much as 250 mV are not at all uncommon on circuit boards.

These (changing) voltage gradients constitute noise, and it is instructive to note that this noise is not coming from somewhere totally outside our circuit. In fact, it is generated by the circuit itself. One guideline for good power supply management is not to focus on preventing noise on the planes from getting into our circuits. It is, instead, to *keep the noise that is generated by our circuits from getting onto the planes in the first place.*

One reason engineers use different regulated power supply regions for different circuits and different stages of the same circuit is to try to prevent noise generated by one (part of a) circuit to interact with, and interfere with, signals in another (part of the) circuit.

NEED FOR POWER PLANES

We have now covered several topics in this book about why power and ground planes are required in high-speed design applications. Here is a summary of some of those important reasons.

Impedance control (Chapters 10, 11, 14): If we want to control trace impedance as a strategy for the control of reflections (using proper trace termination techniques), then good, solid, continuous planes are almost always required. It is very difficult to control trace impedance without the use of planes.

Loop areas (Chapter 9): Loop area can be visualized as the area defined by the path of the signal (traveling down a trace) and its return current. When the return signal is on the plane immediately under the trace, loop area is minimized. Since EMI is directly related to loop area, EMI is minimized when good, solid, continuous planes exist under traces.

Crosstalk (Chapters 12, 13): The two most practical ways to control crosstalk are (a) separation between traces and (b) closeness of the traces to their reference planes. Crosstalk is inversely proportional to the square of the distance between the traces and their reference planes.

Need for Plane Pairs

Decoupling (Chapter 15): The capacitance formed by the proximity of two planes placed close together can be very important and beneficial in circuit decoupling at very high frequencies, where bypass capacitors and their associated mounting and lead inductance begin to have problems.

EMI (Chapter 9): Planar capacitance can be effective in controlling EMI radiations caused by both differential mode and common mode noise signals.

STRATEGIES FOR DESIGNING WITH PLANES

For all these reasons, the use of planes is very important and beneficial in PCB design. But then some relevant questions are these: How many planes should I use? What should be on them? Where should they be placed?

For example, every different power supply voltage is typically distributed on its own plane. It is logical that each different regulated supply of the same voltage also gets its own plane; otherwise some of the regulated supply sections would simply be shorted out. Very often, however, all the different regulated supplies are *referenced* to the same voltage potential—zero volts. Is it possible, then, to have only one ground plane (at zero volts) that can service all the individual regulated power supplies, or do we need a separate ground plane for each individual regulated supply?

In looking at the question of whether each power plane needs its own, separate individual zero-voltage reference plane, we have to look at why the separate power supply exists at all. For example, suppose the overriding issue is power dissipation. Then it may be perfectly fine for two supplies to reference the same ground plane. In some cases, a circuit simply involves two types of ICs with different supply requirements, and a single reference (zero-voltage) plane may work well here, too. And, if we have mixed signals (as in an A/D circuit) but all the circuits are known to be quiet, so noise is not a concern, a single reference plane might be adequate.

As noted earlier, however, one reason for using different regulated power supplies is for noise control—for example, keeping transmitter noise out of a receiver section or keeping digital noise out of an analog section. All the reasons for using planes have a single common denominator—noise control. So it is almost axiomatic in high-speed designs that noise control is the predominant design issue, and our design strategies are undertaken with noise control as an important focus.

Now, if circuit noise is our primary focus, then we almost always need to define separate reference (ground or zero-voltage) planes for each individual regulated supply section. Consider the implications for not doing so. Assume, in a transmitter section, we send a signal down the trace shown in Figure 16-2a. The return signal, although primarily underneath the trace, extends for some distance beyond the edges of the trace. If any receiver circuitry is anywhere near this return gradient, some of this transmitter noise may couple into the receiver section. In addition, there are current gradients that flow from the transmitter-regulated power supply across the ground plane while charging (and discharging) any bypass capacitors that may exist. These currents may also couple into receiver circuitry. The whole point of having a separate receiver power supply section is to try to isolate the two sections. A single ground (reference plane) tends to work against this objective. So, good design practice is to have separate ground planes for the receiver and transmitter sections (Figure 16-2b).

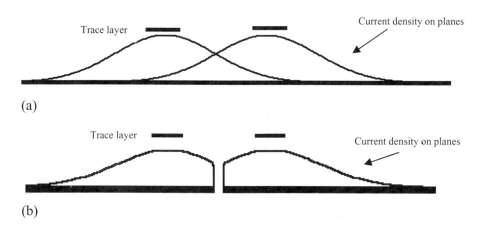

Figure 16-2 Current gradients from one trace can extend across the plane and underneath another unrelated trace (a). Separate ground planes (b) can constrain these current gradients.

Similarly, a digital signal flowing down a trace generates a return gradient on the reference plane under the trace. Such a signal may cause a noise signal to be injected into a nearby analog trace. Again, for this reason, good design practice dictates using separate analog and digital ground planes.

In fact, whenever we use separate regulated supplies for noise control purposes, it is good design strategy to use a separate reference plane for each one, even though all the reference planes are at the same nominal voltage potential—zero volts. Not doing so tends to defeat the very purpose of having separate regulated supplies in the first place.

In summary, separate regulated power supply stages can exist for a variety of reasons. Very often they exist for the purpose of controlling circuit noise, which, in turn, almost always requires the use of planes. If each different regulated power supply section requires its own separate distribution plane, then each should have its own individual (zero-voltage) reference plane, also.

Some Design Rules

Now, based on the preceding, let's assume we have several regulated power supplies (associated with separate circuits on the board), each distributed on its own plane, and each with its own, identifiable, separate reference (zero-voltage) plane. What PCB design strategies and rules are appropriate?

Connecting Reference Planes Together Ultimately, almost all regulated power supplies reference to the same thing, commonly a zero reference (zero-voltage) point in the system. Typically, if one puts an Ohmmeter between the various reference planes, we find that they are all connected together. It is usually important, however, that when reference planes are connected together, the connection is made at a *single point.*

Suppose, for example, this were not true. Suppose we had both a digital and an analog section on our board, and the digital and analog reference planes were connected at two (or more) points. There are conditions under which certain return signals could travel on both planes (not just on the plane under the trace), thereby negating the separation we were trying to accomplish in the first place. Even worse, it is possible (under certain conditions) for noise signals to circulate in a loop around through both reference planes, crossing between them at the two connection points. Such current loops are called *ground loops*. Their origin (cause) is often obscure, but their effects usually are not. The effects include noise problems, EMI radiation problems, and in extreme cases power dissipation and heat problems.

Control over ground loops is relatively simple. If there is only a single connection point for the reference planes, there is no loop over which a signal can travel.

But this poses a new interesting question: Where should that single point be? On some systems, particularly where there are mother and daughter boards, the planes are separately routed back in a "star" fashion to a single point, usually at the primary power supply for the system. If there are multiple regulated power "islands" on a board, these might be connected to each other at single points with zero-ohm resistors (really just a jumper) or with ferrite beads. The beads, of course, are used to block higher frequency components while still allowing a DC connection between the planes. Some argue, however, that if we have been really effective in isolating our power requirements, there are no higher frequency components to filter, and therefore a zero-ohm resistor is all that is necessary.

In A/D circuits, it is common to have digital power and ground planes on the digital side of the IC and analog power and ground planes on the analog side of the IC. We often connect the analog and digital grounds together at a single point directly under the IC (or at least very close to the IC) with zero-ohm resistors or with ferrite beads.

Overlapping Planes If we have separate regulated power supplies with their own reference planes, it is good design practice not to let unrelated portions of the planes overlap. For example, do not let a portion of an analog power plane overlap a portion of a digital ground plane (see Figure 16-3). Remember, a capacitor in its simplest form is simply two conducting surfaces separated by a dielectric. The area over which two planes overlap forms a small capacitor. It may be a very small capacitor, to be sure. Nevertheless, any capacitance provides a path over which noise may travel from one regulated supply to another, working to defeat the purpose for which the separation existed in the first place.

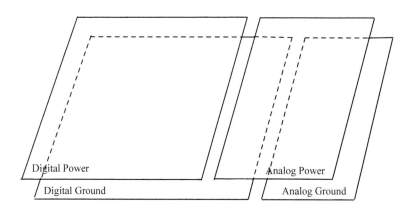

Figure 16-3 Overlapping planes can cause capacitive coupling between unrelated planes.

An important part of the PCB placement process is to place components in such a way that their common regulated supplies (and grounds) can be efficiently grouped together and not overlap other circuits.

Decoupling to Wrong Plane We use bypass capacitors to decouple our circuits—that is, to connect the power and ground planes together (from an AC standpoint) at specific points on the board. It is probably obvious that we don't want to drop a bypass capacitor from one power plane to an unrelated reference (ground) plane. Again, the reason is that we can (and almost certainly will, in this case) couple noise from one regulated supply section into the other. Unfortunately, this mistake can occur accidentally, sometimes fairly easily. As designers, we must be careful to check our engineers' net lists to ensure that this mistake hasn't happened. Even worse, we must guard against making this mistake ourselves. It is very embarrassing when this happens. There is simply no good answer to the question, "Why did you connect it that way?"

Signals Crossing Separations Remember that signals reference to their (usually) nearest plane, be it power or ground. If we have a separation between two reference planes, we never want to route a trace across that separation. Figure 16-4 illustrates an example of routing a digital trace across an analog plane. There are four primary problems that can result from doing this, all of them bad. They parallel three of the reasons we mentioned earlier for using planes in the first place.

1. Good impedance control requires continuous control over geometry, and a continuous return path underneath the trace. If the trace crosses the boundary between two planes, the return signal cannot "jump" the gap. (Don't assume you can solve this problem with a bypass cap. Doing so would couple two unrelated planes together.) This causes an impedance discontinuity and a reflection, and therefore a potential noise problem at that point.

2. If the return signal can't jump the gap, it must find some other path. This almost certainly increases the loop area for the signal and therefore the potential for EMI.

3. If the return signal does somehow find its way onto the analog plane, we now have digital noise on the analog plane, defeating the purpose of the separation in the first place.

4. Suppose two traces cross a separation between two planes. Since their return currents cannot jump the gap, these return currents must find another path. Even though the signal traces are separated from each other, their return signal paths might not be separated, and their *return signals* might crosstalk. Thus, when

signals cross plane splits, crosstalk may result even though there is no apparent cause or reason. This type of crosstalk can be very hard to diagnose.

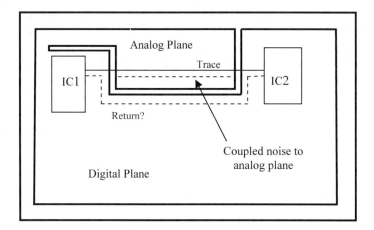

Figure 16-4 Never route a high-speed trace over an unrelated plane.

The primary way to avoid the problem of traces crossing splits in planes is to be careful in the layout and placement stages of the design. Group circuits by regulated supply voltage and then by signal flow. If a particular trace must extend into another regulated supply area, perhaps little "islands" or "peninsulas" (on both the power and ground planes) can be created for the traces involved.

If you cannot find a suitable way of routing traces without violating this design principle, then the circuit design engineer should be consulted. The engineer is the one responsible for deciding whether the design guideline can be relaxed in this particular case, whether the layout should be changed, or whether (in extreme cases) a different component selection is called for.

STACKUPS

We summarized earlier why planes and plane pairs are important in high-speed design. From those rules we can derive two basic rules that help us define what makes a good board stackup.

> *Rule 1:* Every high-speed trace must be referenced to a continuous plane. The reasons include loop area control, impedance control, and crosstalk control.
>
> *Rule 2:* Every power supply should have a parallel power–ground capacitive plane pair. The reasons include help in capacitive decoupling circuits and for EMI control.

A good board stackup meets these two requirements. Any stackup that does not meet *both* these requirements may suffer high-speed problems as a result.

Figure 16-5 was adapted from one created by Rick Hartley.[1] It shows illustrations of good and poor stackups from a high-speed design standpoint. Every "good" design is characterized by every trace being referenced to a nearby plane and every power supply having a power–ground planar pair. Every "poor" design is missing at least one of these two elements.

One conclusion that is drawn from Figure 16-5 is that it is hard to have an efficient six-layer design. That is true. Six-layer designs with four trace layers cannot meet both criteria for effective high-speed design layout at the same time. Therefore, if six-layer boards are going to be used with four trace layers in high-speed designs, we need to understand and accept that some compromises are being made.

---

[1] Rick Hartley, "Controlling Radiated EMI Through PCB Stackup," *Printed Circuit Design*, July, 2000, p. 16.

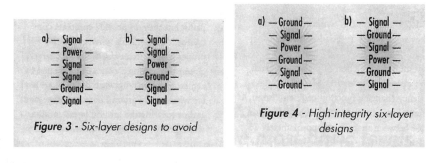
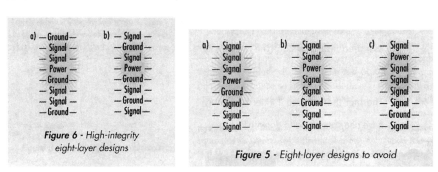

Figure 16-5 Examples of good and not-so-good board stackups (courtesy Rick Hartley and *Printed Circuit Design* magazine).

CONCLUSION

It is common to have different regulated supply voltages on a circuit board. If circuit noise control is not a major issue, then it is possible for all these supplies to reference a common (zero-volt) ground. If circuit noise—especially high-speed circuit noise—is an issue, however, then the regulated supply voltages are commonly distributed on their own individual planes. In this case it is normal for each regulated supply "plane" to have its own individual (zero-voltage) reference plane. These planes are usually constructed as plane pairs on the board. The boundaries defining individual planes are often called *splits*.

Certain, relatively simple design rules are followed when we have split planes:

- Don't allow nonrelated plane areas to overlap.
- Connect reference planes together at only a single point.
- Don't route signal traces across a split or across an unrelated plane.

17 LOSSY LINES AND EYE DIAGRAMS

In days gone by we considered traces on PCBs to be (almost) perfect conductors. On rare occasions we considered their resistance and maybe propagation time, but usually not. Traces almost never contributed to signal integrity problems.

Then, as rise times shortened, we began to worry about signal reflections. In Chapter 10 we learned that signal integrity problems caused by reflections could be managed if we designed our traces to look like transmission lines and if we addressed questions like trace termination. Instead of traces resembling perfect conductors, traces resembled distributed networks of LC components. But if we could measure an effective input impedance whose value is given by Equation 10-1

$$Z = \sqrt{L/C}$$

we could terminate traces in components with the same impedance and still control the reflections.

However, rise times continue to shorten and now new issues are coming into play. Their solution will not be the purview of PCB designers, but designers will need to understand the issues if they are going to continue to grow in their profession.

LOSSY LINES

A transmission line that we consider to be infinitely long and whose input impedance is given by Equation 10-1 is considered to be lossless. That is, whatever signal goes into the front of it arrives at the far end without any distortion. This is not a bad model for well-designed lines at lower frequencies. Although nothing can be truly "lossless," lines can be designed that are generally close enough for most purposes.

But there *are* losses, and as frequencies (or frequency harmonics) get high enough, these losses can become significant and can begin having negative effects on our boards. There are two general areas of losses we need to be concerned about. Both are beyond the scope of this book, so they are simply summarized here so we can illustrate some of their effects. That way, if your circuit design engineers begin setting board design parameters or designing complex termination strategies you will have some understanding of what they are doing and why they are doing it.

Skin Effect Perhaps you have heard about something called the *skin effect.* This describes the tendency of the current to travel along the outside skin of the trace or wire instead of being evenly distributed throughout the entire cross-section of the wire.

The explanation for why this is so is complicated and subtle. If you have read and understood Appendix B you learned that when a current changes rapidly in a wire it (a) generates a changing magnetic field around the wire, which (b) generates a current in the opposite direction in the wire.

This opposing current is what causes the "inertia" we talk about when we describe inductance. At the very first instant, the reverse-generated current is the same as the forward-generating current, and no current flows. But soon the forward current overcomes the reverse-generated current and it does begin to flow. The resistance to the first increments of current creates what we call inductance.

Now consider a wire with a circular cross-section, as shown in Figure 17-1. Assume a current flows evenly through that cross-section and that the current is increasing. The increasing current generates an increasing magnetic field that in turn generates a current back in the opposite direction. Where is that current concentrated?

The strength of the magnetic field is inversely proportional to the square of the distance. Consider a point along the centerline of the wire. All other points that make up the cross-section of the wire are within one radius of the centerline. Now consider another point along the circumference of the wire. More than half the other points that make up the cross-section of the wire are further than one radius away from this point. Therefore the effect of the magnetic field must be *weaker at the circumference* than it is along the centerline.

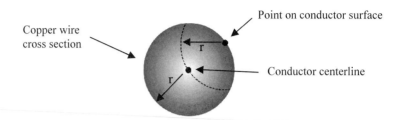

Figure 17-1 Very high-frequency currents will travel on the outside skin of a conductor where the inductive coupling is less (shading suggests current density).

This then means that the *very first* place a sharply increasing current will begin to flow is along the circumference (surface) of the wire, rather than at the centerline (or uniformly across the entire cross-section). At lower frequencies this effect is very transient. Current will quickly spread to the entire cross-section. But consider what happens when the frequency is *very* high and current changes direction *very* quickly. Just when the magnetic field (self-inductance or inertia) is beginning to be overcome, the driving current changes direction. Now we start over again and current begins first to flow on the outer circumference in the opposite direction. Before that current distribution can spread throughout the entire cross-section of the wire, the driving current changes direction again. In this way, the current distribution is always strongest at the outer circumference. In the extreme, when the frequency is high enough, the current distribution never spreads uniformly to the center of the conductor. The higher the frequency, the greater this effect is, and current only flows along a thin portion of the circumference, commonly called the skin of the wire.

Many different applications capitalize on this effect. Transoceanic communication cables, for example, are often made of strong but not necessarily highly conductive steel for strength. The steel cables have a very thin copper (or sometimes even gold) foil wrapped around the circumference. The thin foil is where the current flows. Overall conductivity is not affected because the high-frequency signals are only flowing along the skin anyway.

The skin effect manifests itself as a higher resistance associated with the wire or trace, and the effect increases with frequency. The effective skin depth is inversely related to the square of the frequency. Thus, this is one place where the effective resistance becomes a function of frequency. Since only a fraction of the total cross-section of the wire or trace carries current, and since resistance is inversely related to cross-sectional area, the effective resistance the signal sees increases with skin effect and therefore with frequency.

Dielectric Absorption   We think of dielectric materials as insulators. FR4, for example, is fiberglass. We think of it as passing no current at all on our circuit boards. As very high-frequency currents flow on our traces, however, the very fast-changing currents cause motion at the molecular level of the material. That movement requires energy, which is absorbed from the signal. That, in turn, attenuates the signal. The signal level is reduced, therefore, as energy is absorbed in the dielectric material.

The net effect is as if there were a resistor across the signal leads loading down the signal. The effect is a function of frequency. So again, we have a situation where losses appear as if we have a frequency-dependent resistor across the signal leads.

LOSSY LINE MODEL

In Chapter 10 we illustrated the model of a (lossless) transmission line, shown again here as Figure 17-2:

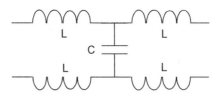

Figure 17-2 A lossless transmission line model.

When we add the "lossy" components to the model it becomes Figure 17-3.

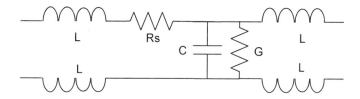

Figure 17-3 A lossy transmission line model.

In Figure 17-3, the Rs represents the (frequency-dependent) skin effect losses and G represents the (frequency-dependent) conductance associated with the dielectric absorption.[1] We use conductance rather than resistance in the model because it simplifies some of the arithmetic. As frequency increases both Rs and G increase. Thus, the higher frequency harmonics of the signal are attenuated.

One significant consequence of losses in the line is that the characteristic impedance of the line changes with frequency. We noted in Chapter 10 that the characteristic impedance of a lossless transmission line is given by the formula in Equation 10-1:

$$Zo = \sqrt{\frac{L}{C}}$$

The characteristic impedance of a lossy line is given by the formula shown in Equation 17-1:

$$Zo = \sqrt{\frac{Rs + j\omega L}{G + j\omega C}} \qquad [17\text{-}1]$$

The frequency range at which lossy lines become an issue is, of course, dependent on many factors. In general, loss effects can become measurable in the 100 MHz range and can become design issues in the 1.0 GHz range.

---

[1] Recall from Chapter 5 that conductance, G, is the inverse of resistance, R.

Eye Diagrams

Eye diagrams are an effective way to visualize the effects caused by lossy lines. But first we need to know what eye diagrams are and how they are created. Start with Figure 17-4.

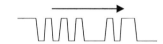

Figure 17-4 Typical data string.

Figure 17-4 illustrates a typical data string of bits with value 1 (high) or 0 (low). If we were looking at this on an oscilloscope screen we would see the right-hand bit first. Then it would shift to the right (in time) and we would see the next bit. So if we were going to read the data string as it passed by a point in our circuit we would read (from right to left) 0-0-1-1-0-1-0-0, and so on.

More generally, if we looked at a specific point in a circuit and watched a large number of data streams passing by that point, and looked only at the three most recent bits, there are only eight possible combinations of bits that would pass by. Figure 17-5 identifies them. In Figure 17-5 the numerical codes represent the data bits as they move past a point in the direction shown. Thus we "see" them in order from right to left.

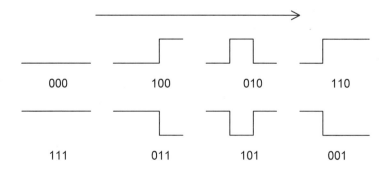

Figure 17-5 Eight possible combinations of a three-bit stream as they pass by a specific point in our circuit. (Read data stream from right to left.)

Now if we took a time exposure of a large number of data bits passing a point, and superimposed them on a single graph, they would look like Figure 17-6.

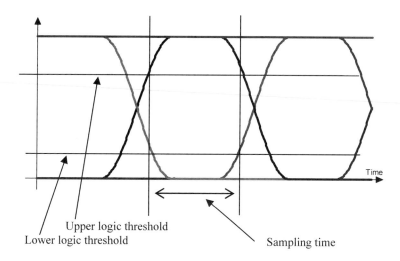

Figure 17-6 All possible combinations, superimposed, of three data bits passing by a point in a circuit.

Figure 17-6 represents an ideal eye diagram. It shows all possible combinations of three data bits passing by a point in a circuit. The transition between logic states is very clean and predictable. The upper and lower logic threshold levels are indicated in the figure. The vertical lines represent the sampling time during which there will be no logic errors. That is, if the data stream is sampled between these times, the data will always be in a correct and valid state. This is referred to as an eye diagram because of the obvious resemblance of the open area to an eye.

When there are losses along a line, the net effect is to attenuate (reduce in amplitude) the higher frequency harmonics. This can cause distortion in the signal. Depending on circumstances, the distortion can range from relatively minor to crippling, causing the system to simply not function. HyperLynx introduced a new version of its LineSim software package in 2003 that allows simulation of lossy lines and eye diagrams. Figures 17-7 and 17-8 illustrate a simulation of a signal at the receiving end of a 20-in. transmission line under ideal and lossy assumptions. This entire area is so complex that I offer these simulations only as indications of how lossy lines may behave, not as suggestions regarding what to do about them.

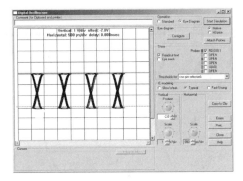

Figure 17-7 Eye diagram simulation for a properly terminated 20-in. transmission line.

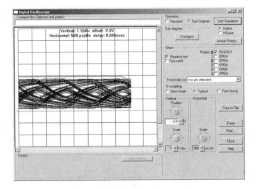

Figure 17-8 Eye diagram for the simulation in Figure 17-7 with significant lossy assumptions.

EQUALIZATION

If the lossy characteristic of a transmission line means that higher-frequency harmonics are attenuated, then we may be able to compensate for that with high-frequency equalization. By that we mean adding active circuits that pre-amplify higher frequency harmonics or passive circuits that act as high-pass filters, attenuating the lower frequency harmonics. Much research is under way at the present time regarding how to do that.

Passive Equalization Passive high-pass filters are a relatively simple way to try to compensate for losses along a line. They consist of simple parallel RC circuits. As frequencies increase, the current passes through an RC equalizing filter with little attenuation, but lower frequency harmonics are attenuated somewhat. Since this is exactly the opposite of what happens along the lossy line, a properly constructed equalizing filter can (at least in theory) exactly compensate for the losses in the line.

Such RC filters can be placed at the beginning of the line (much like a series terminating resistor is), at the end of the line, or can even be built into the receiving device. Figure 17-9 shows what happens with the simulation of Figure 17-8 when a simple parallel RC filter circuit is placed at the beginning of the line. The compensation is not perfect, but the improvement in the signal is evident.

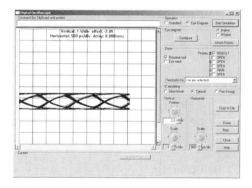

Figure 17-9 RC equalization at the driving end of the line can improve the eye diagram for the simulation in Figure 17-8.

Active Equalization Suppose we have a bit string with a pattern that looks like 0-1-0-1. Here there is a rapid transition from a logical zero to a logical one and back to a logical zero. This represents a very high-frequency sequence traveling down the line, and the signal transitions may be attenuated if the line is lossy. Compare that to a string like 0-1-1-1. There is a rapid transition from the first logical zero to the first logical one, but then the signal stays at the logical one level. This represents a higher frequency transition followed by a lower frequency stabilization.

Let's define the following strategy:

1. Assume the signal transitions between zero and one.
2. Every time a signal transitions from one state to another, we boost its magnitude by .5.
3. If the next clock pulse involves no transition, we boost the amplitude of the signal by .25.
4. After that we will not alter the signal until there is another transition.

This kind of equalization (compensation) can be done in the circuit at the chip level. If done that way, it is called *active precomp*. Figure 17-10a illustrates a bit string without compensation and Figure 17-10b illustrates the same bit string with active precomp. Notice how the active precomp acts to increase the higher frequency parts of the bit string, compensating for the later attenuation that will occur as the signal passes down the line.

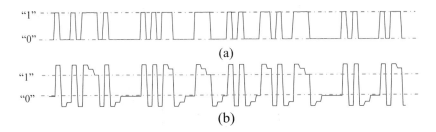

Figure 17-10 Bit streams (a) without any compensation and (b) with high-frequency precompensation. Note that rapid transitions in the signal are amplified to compensate for subsequent losses in the transmission line.

SUMMARY

Lossy line analysis and equalization is going to be a rapidly growing area in the next several years. It will impact device manufacturers, many of whom are already experimenting with or actually designing in and supplying equalization circuits in their products. Passive circuit modules are appearing for both precomp and postcomp placement along lossy lines. Simulation tools will be adding lossy line simulation capabilities to their arsenals.

All of this means that board designers will have to become more conversant with the topic. Board designers will not be responsible themselves for designing equalization circuits for the applications, but they will have to understand them. As we learn more, there are certain to be placement and design rules for equalization circuits just as there now are placement and design rules for transmission line termination circuits.

PART 3

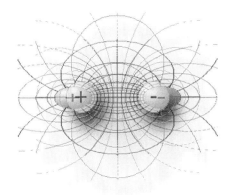

APPENDICES, GLOSSARY, and INDEX

If I covered everything that interested me about the subject of signal integrity and circuit boards, I would never have been able to finish this book. The appendices included in Part 3 assemble several different types of information helpful to the reader. Included here is information regarding how to obtain demonstration files and freeware calculators; tutorials on why inductors induct, on logarithms, and on complex algebra; and a description on how TDRs and VNAs work and why they are useful to us. The last appendix, J, discusses one of my favorite topics—right-angle corners.

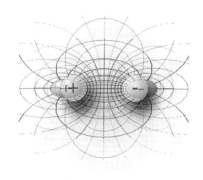

A UltraCAD'S SQUARE WAVE SIMULATOR

*T*his simulation (see Figure A-1) is referenced in Chapter 1. Download the file *www.ultracad.com/prenticehall/square.zip* to a convenient place on your hard drive and unzip it. Execute the resulting square.exe file.

A square wave does not exist naturally in nature. We can think of a square wave as actually being formed by, or made up of, a series of natural cosine waves following the relationship:

$$\cos(\omega t) - \frac{\cos(3\omega t)}{3} + \frac{\cos(5\omega t)}{5} - \frac{\cos(7\omega t)}{7} + \ldots \text{ etc.}$$

where ω is the fundamental frequency of the wave and t is time. The subsequent terms in the relationship contain harmonics. That is, if the fundamental frequency, ω, equals 1.0 kHz (1,000 Hz), then 3.0 kHz (3,000 Hz) is the third harmonic of the fundamental. This is important because if our circuit is going to faithfully deal with the shape of the signal (say a square clock pulse going down a trace) it must be able to faithfully deal with all the different harmonics that are contained within it. This can actually be a very large number.

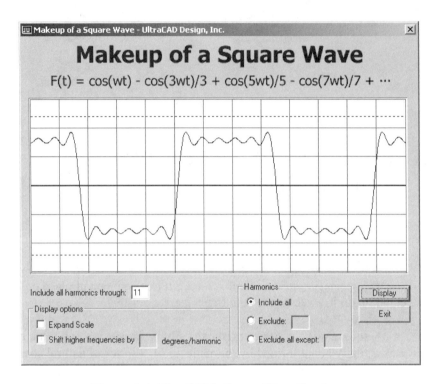

Figure A-1 UltraCAD's Square Wave Simulator.

This simulation can suggest what happens if the circuit does not pass all of the harmonics equally well, and helps you visually understand the makeup of a square wave. Start by entering a number for the "Include all harmonics through:" box. The number 11 is a reasonable place to start. Then press Display. The graph shows the fundamental frequency and the first eleven harmonics following that. (Note that only the odd harmonics enter into this relationship, so the first eleven harmonics only add five harmonics above the fundamental to the equation.) The curve looks like a square wave, but it is not perfect.

Change the number of harmonics to see the effect on the shape of the curve. Determine what you feel is an acceptable approximation to a square wave and see how many harmonics are included. It is that many harmonics the system must be able to deal with, without distortion, to produce the curve you think is acceptable.

This simulation allows you to experiment in some other ways, too. For example, select some number of harmonics and look at the waveform. Then select the

Appendix A

Exclude button and enter the number of a harmonic to exclude from the equation. Then select Display. Note how the waveform distorts. This type of thing could happen, for example, if somehow a filter existed in the circuit at the frequency you just selected that acted to "kill" that particular harmonic. That filter could happen inadvertently, or it could exist as a result of parasitics that no one knew about or expected. This is an insidious problem that can be very difficult to diagnose. (Note that if you exclude an even harmonic nothing happens because even harmonics are not contained in the series relationship.)

If you select the Exclude all except option, you can look at the relative magnitude of any individual harmonic.

The Expand Scale option allows you to see a close-up view of the rise time of the waveform. It expands the display in the approximate area of the fifth vertical grid line. This allows you to see a close-up view of any distortion or ringing associated with the rising edge.

The Shift higher frequencies by option will add an increasing phase shift to each harmonic. Suppose the circuit has some capacitive loading. The capacitive loading can cause a negative voltage phase shift (voltage lags current). If you enter a negative number in the box, each harmonic will be shifted by a multiple of this phase shift. Try these numbers: Enter 99 for the number of harmonics to include and –4 for the shift. Then select Display. The display looks almost exactly like the ringing that is common in many circuits.

Caution: The Shift higher frequency by option does not really reflect real-world conditions. I have included it to *suggest* some effects, but a real-world phase shift would be much more complicated than this. Treat this particular feature as interesting, but not necessarily representative.

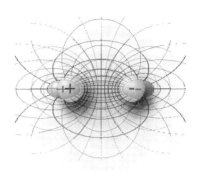

B   WHY INDUCTORS INDUCT

ELECTROMAGNETIC CONCEPTS

*T*here were four people who were particularly important in the development of electromagnetic theory. Their groundwork was crucial to the ultimate understanding of this complex topic.

First, Charles Augustin de Coulomb has been credited with Coulomb's Law (1785) in which he stated that there are two kinds of charge, positive and negative. Like charges repel, and unlike charges attract, with forces proportional to the product of their charge and inversely proportional to the square of their distance.

Carl Friedrich Gauss is generally credited with stating that every magnetic pole is actually part of a dipole that includes an equal and opposite pole. One pole cannot exist without its opposite. Magnetic force is applied along a vector with a direction that is a line along which the force acts. Magnetic force is inversely proportional to the square of the distance.

André Marie Ampere is credited with Ampere's Law: An electrical current is accompanied by a magnetic field with a direction at right angles to the direction of the current flow.

Finally, in 1831, Michael Faraday published Faraday's Law of Magnetic Induction: A *changing* magnetic field is accompanied by an electric field that is at right angles to the change of the magnetic field.

These people, over a 50-year period, established some very important groundwork, but they were all (particularly Faraday) experimental scientists, not great mathematicians. It was James Clerk Maxwell who combined these important observations and laws into a single set of equations that comprised a closed system. By that, we mean that every phenomenon and every relationship is self-contained in the set of equations, and any change in any one variable will flow through to every other variable in the system. There are no outside variables or forces needed or allowed.

This set of equations has become known as Maxwell's Equations, and they were published in his famous <u>Treatise on Electricity and Magnetism</u> in 1873. Maxwell is often treated as though he developed all the theory behind the equations, which is not true. For almost 100 years before him, people were developing significant parts of what would become known as electromagnetic theory. Maxwell's considerable contribution was seeing how to consolidate the work that several others had done earlier into a single, closed set of equations. That was no small feat, and Maxwell's Equations are still, today, crucial to our work and understanding in this area.

If you are wondering why I am going to spend a little time on electromagnetic theory, let me give you two reasons. First, it is at the heart of inductance. Inductance (and related issues like crosstalk) is an electromagnetic effect. Second, the first two letters of EMI stand for "electromagnetic." We need an understanding of this theory to help us understand destructive EMI effects.

RIGHT-HAND RULE

Let's begin our discussion by looking at Ampere's Law. If a current flows through a wire, there will be a magnetic field around the wire (Figure B-1). The strength of the magnetic field is directly related to the amount of current that flows. Small currents have weaker magnetic fields; larger currents have stronger magnetic fields. We can find the direction of this magnetic field by the right-hand rule. Point the thumb of your right hand in the direction of the current flow. The magnetic field force curls around the wire in the direction in which your fingers curl.

Appendix B   331

Figure B-1 Right-hand rule.

      This effect can be easily demonstrated at home using the famous compass experiment (Figure B-2). Connect a wire to a battery and place it under a compass so that the current flows toward the North Pole. Point your thumb in the direction in which the current flows (north). Notice how your fingers curl around the wire. The magnetic force around the wire, and above the wire (closest to the compass), will be in an easterly direction. When the wire is connected to the battery, the compass needle will swing toward the east, aligning with the magnetic force around the wire caused by the current flow. Disconnect the wire from the battery and the compass needle will point back toward the north.

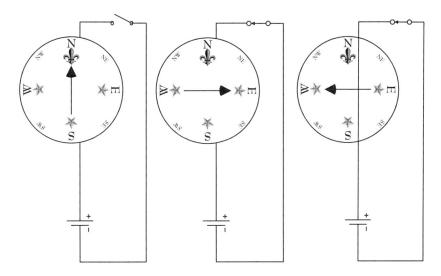

Figure B-2 The classic compass experiment.

If you now place the wire above the compass, notice how your fingers below the wire (closest to the compass) curl toward the west. When the wire is connected to the battery, the compass needle will swing to the west, aligning now with the magnetic force under the wire. Disconnect the wire from the battery again and the compass needle will swing back toward the north.

This simple experiment confirms the absolute relationship that exists between electric current and magnetic fields. Remember that current is the flow of electrons, and it is the magnetic property associated with electrons that is at the heart of this relationship.

COIL

We can strengthen and focus the magnetic lines of force if we coil the wire (Figure B-3). Note that the current through the wire in Figure B-3 is flowing from bottom to top. If, at any point along the wire, we point the thumb of our right hand in the direction of the current flow, our fingers curl upward inside the coil and downward outside the coil. The magnetic lines of force tend to be focused inside the coil, especially if the coils are close together. Since our fingers curl upward inside the coil, the north pole of the magnet will be at the top of the coil.

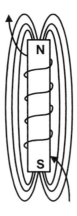

Figure B-3 Wire coils help "focus" the magnetic field.

Appendix B

The strength of the magnet will be increased if we (a) increase the current, (b) add more turns to the coil, or (c) add a ferrite core inside the coils to further focus the magnetic force.

SIMPLE MOTOR

In Figure B-3 we have taken a coil and passed a current through it to create an electromagnet. If we now place that coil between two other magnets, as shown in Figure B-4, we have the makings of a simple motor. The north pole of our coil will be repelled by the magnet with the north pole facing it, and it will be attracted to the magnet with the south pole facing it. Thus, there will be (in Figure B-4) a clockwise torque acting in the coil.

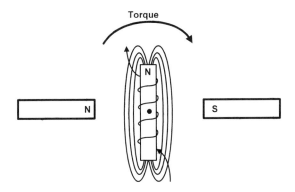

Figure B-4 Simple motor.

This is, in fact, the theory and design behind a basic electric motor. Real motors are more complicated in their design, of course. There are more complex mechanisms for getting current to the coil and for creating the internal magnetic fields, but the basic principle behind any electrical motor is as shown in Figure B-4.

If you look closely at the heart of the meter movement for a basic analog electrical meter, it will look much like Figure B-5. Current through the coil creates a torque that tries to rotate the coil in a clockwise direction. A spring pulls the coil back. The degree of rotation will be the point where the force created by the magnetic field just balances the restraining force of the spring. Look closely at the meter movement of a typical analog Volt-Ohm meter (sometimes called a volt-ohm-milliammeter, or simply VOM) and you can see the magnet assembly, the coil, and the restraining spring.

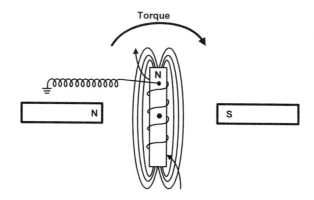

Figure B-5 Fundamental concept of an electric analog meter.

SIMPLE GENERATOR

Suppose we now take our coil and place another magnet beside it, as shown in Figure B-6. If the magnet is not moving, nothing much happens. Faraday's Law states that a changing magnetic field is accompanied by an electrical field, but says nothing about a stationary magnetic field. If we move the magnet toward the coil, however, the magnetic lines of force around the magnet cut across the coil's wires as the magnet moves. Thus, the wires see a changing magnetic field, and an electrical current is induced in the coil. The magnitude of the induced current is related to (a) the strength of the magnetic field, and (b) how fast the field is changing. Faster changing, stronger fields induce stronger electrical currents.

There are some important aspects here. First and foremost, the magnetic field must be changing. This concept is often difficult for people to grasp when they are studying this for the first time. If the magnet is a bar (iron) magnet, changing magnetic fields are created through motion (as suggested in Figure B-6). A changing magnet field can also be created by passing a changing current though a coil. Either way, something must be *changing* to induce an electrical current in a nearby coil. If everything is stationary, no current will be generated.

Appendix B

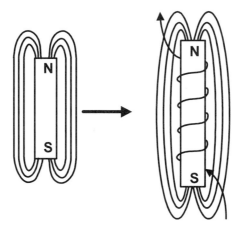

Figure B-6 A moving magnet can generate a current in an adjacent coil.

Looking ahead for a moment, one way to generate a changing magnetic field is with a changing current. The *strength* of the magnetic field is directly related to the magnitude of the current. The *rate of change* of the magnetic field is directly related to how fast the current changes. One of the biggest culprits in high-speed design signal integrity problems is di/dt—how fast current changes per unit time. Here is di/dt in the context of a changing magnet field. A changing current, di/dt, causes a changing magnetic field that can generate an unwanted noise signal in an adjacent wire.

Referring back to Figure B-6, it can be instructive to ask which way the induced current will flow. Is it going to flow upward through the coil, or downward? The answer can be tricky. The induced current flowing through the coil generates its own magnetic field. According to Lenz's Law (Heinrich Friedrich Emil Lenz), what is determined is the induced magnetic field. The induced magnetic field will be generated so that it will oppose the changing magnetic field generating the electrical current in the first place. We can then determine the direction of the induced current through the right-hand rule. Curl your fingers in the direction in which the induced magnetic field must be oriented; your thumb will then point in the direction of the induced current.

In Figure B-6, the moving magnet has a north pole at the top and it is moving toward the coil. The induced magnetic field in the coil will be generated with a north pole at the top of the coil to repel the approaching magnet. Thus the electrical current will flow in whatever direction is necessary to generate a magnetic north pole at the top of the coil. In Figure B-6 this direction is from bottom to top in the coil.

In Figure B-7 the coil is wound in the opposite direction. The induced north pole still must be at the top of the coil, so in this case the induced current must flow from top to bottom in the coil to generate a north pole at the top.

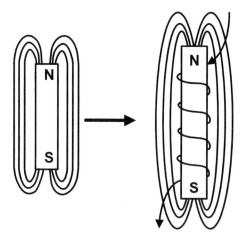

Figure B-7 The direction in which the current flows depends on how the wire is coiled.

If the magnet moves away from the coil, the situation is exactly reversed. The changing magnet field induces a current in the coil, but this time the induced current flows in the opposite direction. That is so the top of the coil becomes a south pole and attracts the moving magnet, trying to stop its motion.

Thus, there is no easy rule to help define in which direction an induced current will flow. The basic rule is that the induced current will flow in a direction so that the magnetic field it generates will counteract the changing magnetic field that induces it in the first place. The actual direction depends on how the coil is wound.

Appendix B

When you think about it, the fact that the induced magnetic field counteracts the inducing field must be so. If it were the other way around, everything would be reinforcing. A moving magnet would generate a current to attract the magnet faster, which would induce a larger current, and so on. A runaway situation would result. We know intuitively that this doesn't happen. Induced currents always act to counteract the forces inducing them.

Figures B-6 and B-7 illustrate the basic principles behind a generator. As with motors, the real-world designs are more complicated for reasons of efficiency and manufacturability, but all electrical generators are based on this simple basic concept.

Figure B-4 illustrates a basic motor. But what if we mechanically turned the coil, instead? It would then see a changing magnetic field caused by its motion within and relative to the magnetic field from the stationary magnets. Thus, if we mechanically rotated the coil in Figure B-4, a current would be induced (generated) in its coil. In that case, Figure B-4 would illustrate a generator.

If Figure B-4 can illustrate both a motor and a generator, what is the difference between them? The difference between a motor and a generator depends on where and how energy is applied. If electrical energy is applied and motion is generated, we have a motor. If motion is applied and electrical current is generated, we have a generator. The same device can be both a motor and a generator, depending on how it is used.

You can demonstrate this for yourself by going to an electronics store and buying a simple 3-volt DC motor. Hooking it up to a battery will confirm that it is a motor. Then connect a sensitive VOM (ammeter) to the terminals of the motor and turn the shaft with your fingers. (An analog meter works best, because the generated signal only exists for the short moment that you turn the shaft. Remember, only *changing* magnetic fields generate an electrical force.) You will see the ammeter needle jump momentarily as your fingers turn the shaft, confirming that you are generating a signal.

A car has an electrical starter motor and also a separate electrical generator (alternator). These are different devices (not a single device used for both purposes) because the demands on them are extreme and they each need to be optimized for their specific use. A gasoline-powered golf cart, however, usually has a single device that acts as both a motor and a generator. When you push the gas pedal down, the circuit closes and the battery turns the device (motor) to start the engine.

When the engine starts, it turns the device (generator) faster and generates enough current to recharge the battery connected to it. This is a practical example of a single device that operates as both a motor and a generator in the same application.

INDUCTANCE

Now, let's look at the situation in Figure B-7 but skip the magnet; we'll use another coil, instead (Figure B-8). If we apply a changing current through the coil on the left, it will generate a changing magnetic field that will induce a current in the coil on the right.

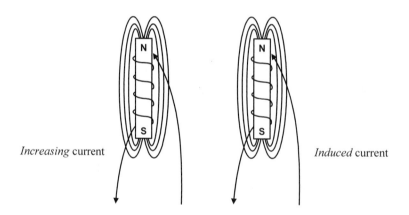

Figure B-8 Very basic transformer action.

This is the principle behind a basic transformer. Current applied to one winding induces a current in another winding.

Figure B-8 illustrates a pretty poor way, actually, to make a transformer. The most obvious improvement would be to wind the two coils (the driven, or primary, and the induced, or secondary) on the same axis, close to each other. There are many other tricks that are beyond the scope of this discussion that people use to make very efficient transformers. Nevertheless, Figure B-8 illustrates a case in

Appendix B

which one winding can (and will) induce a current in an adjacent winding, which is the definition of a transformer.

Now if Figure B-8 illustrates how one coil can induce a current in an adjacent coil, so does Figure B-9. The only difference between Figures B-8 and B-9 is that we have straightened out the windings into straight wires. As we have seen before (the compass experiment) a current in the wire on the left generates a magnetic field around the wire. If we apply a changing current to the wire on the left, the magnetic field it generates will also be changing. That changing magnetic field will cut across the wire on the right, and will induce a current in the wire on the right. The direction of the induced current will be such as to (a) generate its own magnetic field which will (b) counteract the magnetic field from the wire on the left. Thus, the induced current will be in the opposite direction to the inducing current.

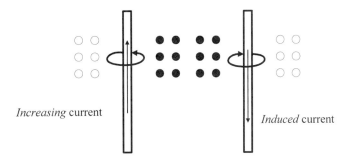

Figure B-9 Two adjacent wires will couple together.

The coupling illustrated in Figure B-9 is pretty weak. It would be stronger for the coils illustrated in Figure B-8 and, of course, much stronger if we had a transformer whose windings were on a common axis. Nevertheless, there will be coupling. The strength of the coupling will be related to (a) how close the wires are to each other, and (b) how strong the changing magnetic field is. The changing magnet field will be stronger for (a) higher currents and (b) higher frequencies. Here is our problem di/dt again. Higher values of di/dt generate stronger, faster changing magnetic fields, which in turn can cause greater induced currents in adjacent wires.

On a PCB we might have traces only 6 mils apart carrying signals with rise times (dt) of a nanosecond. This is close enough, with a high enough di/dt, that coupling can be a problem. We know this type of coupling as crosstalk.

This type of coupling is called *mutual inductance*. A current in one wire is affecting the current in an adjacent wire. Now let's carry this discussion one final step (Figure B-10), by removing the adjacent wire.

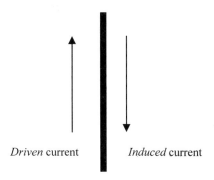

Driven current   *Induced* current

Figure B-10 A single wire will couple into itself.

A changing current in a wire will cause a changing magnetic field that will induce a current *in itself* in a direction that will counteract the changing magnetic field that generates it.

This is called *self-inductance*, or simply inductance. The induced current will be in the opposite direction of the driven current, thereby opposing or "fighting" the driven current. This is the source of the inertia we talked about in Chapter 3. The initial flow of current causes an induced current in the opposite direction, which *for the first instant* results in a net flow of no current at all (until the initial inertia can be overcome). Then, if the change in the magnetic field slows (i.e., the current doesn't change), the induced current begins to reduce and more net current flows. In the steady state case, there is no change in the current, no change in the magnet field, no induced reverse current, and therefore no more inductive effect.

If the driven current is constantly changing (as with an AC sine wave), the magnetic field is constantly changing, and there is a continuing reverse current being generated that impedes the primary current flow. This is the inductive impedance that is equivalent to resistance in the DC case.

It is probably intuitive that the inductance here (Figure B-10) is pretty small. It takes either a lot of current or a rapidly changing magnetic field for this little amount of inductance to have much effect. In high-speed circuits where we are dealing with rise times (dt) in nanoseconds, this small amount of inductance can begin to pose a problem. This is why we in the industry stress rise time, not frequency, as the issue. For the same magnitude current, a square wave will have a significantly faster rise time than will a sine wave, so a square wave will generate faster changing magnetic fields, which in turn will be more susceptible to smaller values of inductance.

You have probably picked up on the idea that the magnitude of the inductance is a function of the configuration. Figure B-11 illustrates some general guidelines as to how inductance might change with configuration (increasing from left to right).

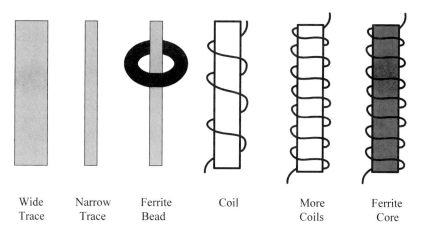

   Wide   Narrow   Ferrite   Coil   More   Ferrite
   Trace   Trace   Bead   Coils   Core

Figure B-11 Methods (left to right) for increasing inductance.

Larger wires have smaller inductance than smaller wires. Think of this as being the result of smaller current densities and smaller magnetic field densities the more wire there is to "spread out" over. On a PCB that is why you sometimes see recommendations that traces between bypass capacitors, for example, and IC pins be as "fat" as possible. That is also why power supply planes generally have lower (smaller) inductance than traces.

Since inductance is a function of length (i.e., we can think of inductance per unit length) shorter wires have less inductance than longer wires. That is why we tend to keep lead lengths and trace lengths as short as possible on PCs.

We can increase inductance at a point by placing a ferrite bead over the wire. Ferrite (iron) has special magnetic properties and helps "focus" magnetic fields. Therefore, the magnetic fields will be stronger around the ferrite bead. Thus, a ferrite bead will increase inductance and its effects (resisting rapidly changing currents) at that point.

A wire shaped as a coil will have higher inductance than a straight wire. The inductance will increase as (a) we increase the number of coils and/or (b) "tighten" the coils (so there are more coils per unit length). Just as a ferrite bead will increase the inductance of a wire, a ferrite core within a coil will increase the inductance of the coil.

Although there are many ways to create capacitance, there are only a few ways to make inductors. For the most part, all fabricated inductors are coils of wire, probably with a ferrite core. When used in power supplies, inductors tend to be pretty big (at least relatively speaking) compared to other types of components. Wirewound inductors tend to have some parasitic capacitance between the windings. Thus, it is much harder to fabricate a "pure" inductor than it is to fabricate a "pure" resistor or capacitor.

For these reasons, circuit designers don't design very many inductors into their circuits. When an inductor might be useful in a design, we often try to find another way to achieve the same result.

There are lots of little inductors on our boards caused simply by lead lengths. These might be traces or they might be component leads. Even the wire-bond leads inside IC packages create little inductors that can be of concern in circuits with fast rise times. These are discussed particularly in Chapter 15 when we talk about ground bounce and the role of bypass capacitors in circuits.

Appendix B 343

RETURN CURRENTS

Many of us have heard that when a trace is routed close to a plane, the return current "wants" to be on the plane directly under the trace. The reason is because that is where the impedance is lowest. And the reason for *that* is mutual inductance.

Consider the two traces shown in Figure B-12. Current i_1 is coupled into Trace 2 with a coupling coefficient k. The coupling coefficient, k, is determined primarily by how close Trace 2 is to Trace 1 (distance D). The coupled current goes in the opposite direction. If Trace 2 carries the return current, then the coupling helps "boost" the return current. The return current, in turn, couples into Trace 1, helping to "boost" current i_1. If the currents are "boosted" (i.e., the traces are closer together), then it takes less energy to move the current along than if the currents are not boosted (i.e., the traces are further apart). If it takes less energy to move the same current, that means the impedance to the current flow must be less. Therefore, the impedance is lower if the traces are closer together than if the traces are further apart.

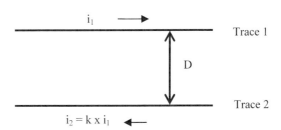

Figure B-12 A changing current in Trace 1 will couple into an adjacent Trace 2.

This is exactly the same phenomenon we were talking about with Figure B-9. Current in one wire can electromagnetically couple to the other through mutual inductance. In the special case where the other wire carries the return current, this coupling actually helps the flow of current by effectively reducing the impedance to the flow.

Now consider Figure B-13. Assume current is flowing down the trace and the return current is flowing back along the plane at point a. The return current will boost the primary current. But if the return current flows back at point b, the boost will be stronger because the distance between the current and its return path will be shorter. If the boost is stronger, the impedance is lower. The point of lowest impedance will be where the boost is strongest (i.e., point c when return current is directly under the trace). And this tendency will be stronger as our friend, di/dt, gets larger—that is for higher speed circuits.

Figure B-13 Signal return currents will be positioned on the plane directly underneath the signal trace.

The fact that the return current wants to flow directly under the trace is also very important in other contexts. It has implications for such things as impedance control (control of reflections), loop areas (EMI radiation), and even crosstalk. As designers, we sometimes do things that prevent the return current from flowing where it wants to. The implications of that are almost always bad.

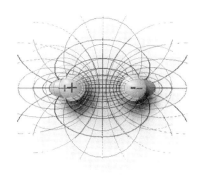

C LOGARITHMS

*T*his appendix covers the definition of logarithms (logs), how we convert from logs of one base to logs of another base, and why we care about using logs in the first place.

DEFINITION

A logarithm is an exponent of a number we call the base of the log. In general, if we have an expression of the form $y = a^x$, we say we take the log (to the base a) of y to obtain x, or $\log_a(y) = x$. This looks kind of abstract, so lets look at the most common form of logarithm called, appropriately enough, a common log:

$$\text{If } y = 10^{.5}, \text{ then}$$
$$\log(y) = \log(10^{.5}) = .5$$

(We omit the subscript 10 in common logs, assuming it is understood.)

In this illustration, the base is 10 and the log is .5. Similarly, the (common) log of 100 (10^2) is 2 and the log of one million (10^6) is 6. A more interesting case is the log of something like 486,397:

$$\log(486397) = 5.687, \text{ since}$$
$$10^{5.687} = 486,397$$

(Note: In all following illustrations there might be slight differences due to rounding. The use of logs with real numbers rarely results in exact results.)

Logs work in reverse, too. For example, if we say the log of a number is something like 3.875 (i.e., log(y) = 3.875), what is the number? This is called an inverse log operation and is written as $\log^{-1}(3.875) = y$. The answer is 7,499, since $10^{3.875} = 7,499.0$.

Before computers and calculators we had to look logs up in tables. Now, virtually every handheld calculator will take a log of a number simply by pushing a button, and an inverse (or anti-) log by pushing two buttons.

Base 10 is the most common log, hence the name. There is another base that we routinely use in electronics, the base e, where e is the natural constant equal to 2.7182818...... To differentiate between the common log and the natural log, we use usually use the name ln() to mean the natural log. So, if we have an expression $y = e^x$, then:

$$\ln(y) = x$$

For example, if $y = e^{7.25}$, then ln(y) = 7.25. Similarly, if we have y = 1,000, then ln(y) = Ln(1,000) = 6.91 (since $e^{6.91}$ = 1,000). And if ln(y) = 6.25, then $y = \ln^{-1}(6.25) = 518$, since $e^{6.25} = 518$.

There are three special log values. Log(1.0) always equals zero, no matter what base we are working in. Any value raised to the zero (0) power is one. The log of zero (log(0)) does not exist in any base. There is no power a number can be raised to whose result is zero. Anytime we take the log to the base n of the value n, the result is one. That is, the log(10) = 1.0, ln(e) = 1.0; even $\log_x(x) = 1.0$.

OTHER BASES

Although base 10 and base e are the most common bases for logarithms, conceptually any base can be used. For any expression $x = a^y$, then $y = \log_a(x)$. For example, we might routinely want to work in base 2 (binary system). Thus, if $y = 2^5 = 32$, then $\log_2(y) = \log_2(32) = 5$. We might also want to work in base 8 (octal system). So $\log_8(247) = 2.65$, since $8^{2.65} = 247$. Or, we might want to work in the hexadecimal (hex) system, base 16. We could even work in a system that makes little intuitive sense, such as 7/8. For example, what is $\log_{7/8}(125)$?

CONVERTING BETWEEN BASES

Although we have tables, and now calculators, for common and natural logs, we generally don't readily have them for any other base. It is not too difficult, however, to convert from any one base to any other base.

To start with, the very first pair of equations we started this appendix with present a special case. The general case is:

$$\text{If } y = 10^{.5}, \text{ then}$$
$$\log(y) = (.5)\log(10)$$

We reduce this to $\log(y) = .5$ since $\log(10) = 1$. But if we were in another base, like base 2, then the correct result would be:

$$\log_2(y) = (.5)\log_2(10)$$

Now, how do we determine $\log(y)$ for any base? For example, what is $\log_{7/8}(125)$?

(1) Let $y = a^{x1}$
(2) Therefore, $\log_a(y) = x1$
(3) Also, $\log_{10}(y) = (x1)\log 10(a)$

In this example, y = 125, a = 7/8, and x1 is the answer we are looking for.

Note that we can now solve for x1:

$$x1 = \log_a(y) = \log_{10}(y)/\log_{10}(a)$$

Therefore, to find the log of y for any base, take the log of y to base 10 and divide by the log (base 10) of the base of interest. In our earlier example,

$$\begin{aligned}\log_{7/8}(125) &= \log_{10}(125)/\log_{10}(7/8)\\ &= 2.0969/(-.05799)\\ &= -36.1586\end{aligned}$$

We can check this by seeing if $(7/8)^{-36.1586} = 125$, and it does, as we can confirm from most calculators or a spreadsheet.

USEFULNESS OF LOGS

So what? Why do we care about logs? That's a fair question. I'll give you three answers in this appendix.

First, before the days of electronic calculators (about 1970 or so) a complex problem could be very difficult to work out. For example, suppose you had to solve for

$$y = (25.875)(36.54^2)(74.36^{1.75})/.8278^{.9}$$

Today, we'd put that in a spreadsheet or grind through it on a calculator. Before that we would simplify it using logs:

$$\log(y) = \log(25.875) + 2\log(36.54) + 1.75\log(74.36) - .9\log(.8278)$$

That may not look too simple, either, but it has been reduced to four log look-ups (in a table), three multiplications, and some addition. When we had a numerical answer on the right side, we'd do an inverse log and solve for y.

So, at least during the days before electronic calculators and computers (i.e., when problems were solved with paper, pencils, and slide rules) logs simplified complex calculations.

A second purpose for logs is to linearize an expression. Suppose you wanted to do a curve fitting analysis and find the best curve that fit the equation:

$$y = Bo(x1^{B1})(x2^{B2})$$

Appendix C

That is, for a set of data x1,x2,y, what is the best guess for the coefficients Bo, B1, and B2? Setting up a least squares curve fitting model for this is not too difficult, but it is nonlinear. It can be converted to an easier linear form using logs:

$$\log(y) = Bo + (B1)\log(x1) + (B2)\log(x2)$$

which looks more like the familiar

$$y = bo + b1(x1) + b2(x2).$$

This form may be easier for doing a least squares fit for the coefficients Bo, B1, and B2.

We often use logs for scaling graphs. Suppose we want to graph the reactance of a capacitor as a function of frequency (recall $Xc = -1/j\omega C$). If the horizontal scale is linear, it is impossible to look at the relationship in the region of, say, 1 kHz, 10kHz, 100kHz, and 1MHz on the same sheet. If we can see details in the kHz range, then 1 Mhz will be far away (in the next room or block). If we can see detail in the MHz range, then the kHz range will be but a spec.

By converting the graphical scales to logarithmic rather than linear, the graph becomes much easier to read. See Figure C-1. The linear graph has a shorter range than the log one, and still offers much less detail than the log one does.

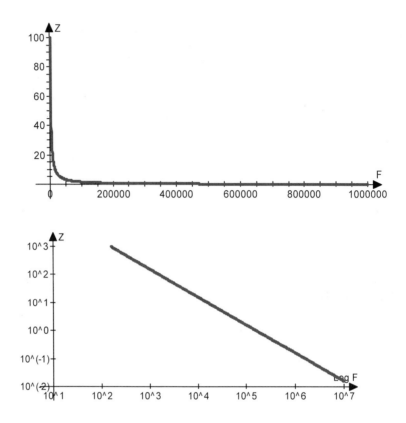

Figure C-1 Reactance of a 1 uF capacitor graphed on linear (top) and log (bottom) scales.

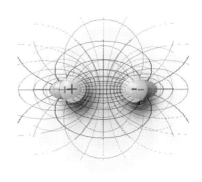

D PHASE SHIFT SIMULATION

Background

*T*his simulation is intended to help you visualize the types of phase shift that occur in circuits as the relationship between circuit components changes. Download rlc.zip from *www.ultracad.com/prenticehall* and copy it to a convenient place on your hard drive. Unzip it to obtain rlc.exe. When you execute this file, the program will start (Figure D-1).

All component values are initially set to 1.0 (Ω, uH, and uF, respectively). The angular frequency, ω, is fixed at 10^6. You will note that this is the resonant frequency under these conditions. Component values can be changed by the sliders across the top and the values are shown in the text boxes at the right. The total impedance and the phase angle are also shown at the right, and the system is at resonance when the components are selected such that the phase angle is zero.

The schematic of the basic series RLC circuit is shown at the left. The circuit simulation can be set up for either constant current input or constant voltage input by the selection buttons at the lower right of the window. The check boxes across the bottom select what is displayed. The options include the voltage across the resistor (V(R), green), the inductor (V(L), yellow), and/or the capacitor (V(C), blue). The input current ((I(in) in mA) and the total voltage (V(in), which is the sum of the other three voltages) can also be displayed. The white curve is always the constant input variable (voltage in or current in) and the red curve is the other input.

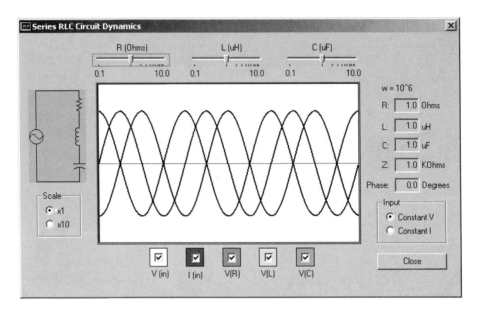

Figure D-1 UltraCAD's Phase Shift Simulator.

This simulation can be complicated to get into, so you might find it convenient to start with the following sequence. After completing this sequence, you can play with the simulation in other ways that interest you.

FIRST STEP

Start with the input set to constant current. Set all component values to unity. Select I(in), V(R), V(L), and V(C) for display. You probably see only three curves, and either the white or green one is missing. Why? Because the voltage across the resistor and the current are always in phase with each other. With these component selections, their magnitudes are also equal. So the curves lay on top of each other. Now deselect the I(in) curve for display.

Note how the V(C) curve lags the V(R) curve by 90 degrees and how the V(L) curve leads the V(R) curve by 90 degrees. Note that these phase relationships *will not change* throughout this entire simulation. The voltages across these three components will always maintain this phase relationship no matter what values the components are set to.

Note how the V(C) and the V(L) curves are exactly equal in magnitude and opposite in phase. Their sum is always zero. That is the definition of resonance. That is why, under these component values, V(in) is exactly equal to V(R).

Change the value of R. The voltage V(R) changes in response. Why doesn't V(L) or V(C) change? Because the input is constant current. Therefore, a change in the value of the resistor doesn't change the current. Therefore the voltage across the other components doesn't change. Reset the value of R back to unity.

Add V(in) as a selection for viewing. Change the value for C. Note several things. The voltage V(C) changes. Therefore V(in) changes, since V(in) is the sum of V(R), V(C) and V(L). The phase relationship changes, since we are no longer at resonance. The V(in) curve shifts in phase relative to I(in). Why doesn't V(R) or V(L) change? Because we are still dealing with constant current in, and since their component values are not changing, the voltage across them doesn't either.

If the phase relationship is changing, why don't the curves shift relative to each other or to the vertical axis? Because we are dealing with constant current in. Therefore, V(R) is always in phase with I(in) (which is constant) and the other two voltages are always 90 degrees shifted in phase from V(R). Reset all component values back to unity.

SECOND STEP

Clear the check boxes for V(R), V(L), and V(C) and select the boxes for V(in) and I(in). If all component values are set exactly to unity, you probably see only one curve, not two. Why? Because when the component values are set to unity, the voltage and the current have the same magnitude. And since the system is at resonance, they have the same phase relationship. Therefore, they superimpose exactly on top of each other.

Now change the value for R. Note how the input voltage changes in proportion to R. Remember that the input current is constant, so less voltage is required for smaller resistance. Note how the phase relationship does not change simply because R is changing. Reset R to unity.

Decrease the value for C a little. Note that the voltage curve shifts to the right and the phase angle display shows a negative number. Why? Because voltage lags current when the circuit is capacitive. Note that the voltage peak occurs later than (lags) the current peak. Decrease the value for the inductor. The voltage curve continues to shift. Move the inductor and capacitor sliders all the way to their minimum positions, and select the x10 button to change the display scale. Note how the voltage curve lags the current curve by almost 90 degrees (the text box probably says something close to −84 degrees). This can be seen by comparing the position along the horizontal (time) axis of the peak of the voltage curve and the zero-crossing of the current curve.

Change the L and C settings to their other extreme. Note that the phase shift changes from almost −90 degrees (lag) to almost +90 degrees (lead).

THIRD STEP

Reset the component values back to unity and set the input condition to constant voltage. You may only see one curve, since with unity values (resonance) the V(in) and I(in) curves will be equal in magnitude and in phase. Note that the V(in) and I(in) curves have changed color. The white (constant) curve now indicates the voltage in.

Change the value for R. Note that input current changes inversely with R (since we now have constant voltage in). The phase relationship doesn't change because the LC circuit is still in resonance.

Change the value for C. Note that the phase relationship changes, and as C decreases current begins to lead voltage (and correspondingly, voltage lags current). Reset all component values back to unity.

FOURTH STEP

Deselect V(in) and I(in) for viewing, and select the other three voltages. Change the value for R. Note that V(R) does not change, but V(C) and V(L) both do. Why? This is a good question. First, with unity component values, V(C) and V(L) are at resonance, so their voltages are equal and opposite and cancel. So V(in) equals V(R). And since V(in) is now constant, so must V(R) be. But as R changes,

Appendix D

I(in) must change. As I(in) changes so do V(C) and V(L) change in response, but they change together, so that they are always equal and opposite and cancel.

Reset R to unity and begin to reduce the value for C. As C gets smaller, the voltage across it, V(C), increases. Since the total voltage is constant, the voltage across V(R) and V(L) must both decrease. That means Z must be increasing so I(in) is decreasing. Watch the values for Z and the phase angle in the text boxes.

The curves shift to the left, but they are all shifting exactly in the same phase. Remember, the phase relationship between V(C), V(L), and V(R) never changes.

This simulation allows you to visualize what is meant by phase shift and to see what it is that is shifting with respect to what else. Experiment with other combinations of component values and input conditions and then try to understand (and explain) what happens.

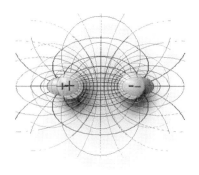

E   COMPLEX ALGEBRA

*T*he use of complex math and imaginary numbers is routine in electronic analysis. This appendix includes some Web links, hints about complex number resources available on PCs, a note on Euler's Formula and how it suggests the relationship between exponential functions and electronic sine waves, and a pretty good article summarizing the topic reprinted from Microsoft's Encarta.

WEB SITES

If you do a search for "imaginary numbers" on the Web you will get a great many hits. Some reasonable places to start are:

http://www.math.toronto.edu/mathnet/answers/imaginary.html ("Do Imaginary Numbers Really Exist?")

http://ca.yahoo.com/Science/Mathematics/Numerical_Analysis/Numbers/Complex_and_Imaginary_Numbers

http://forum.swarthmore.edu/dr.math/ (Note: This site has good links for many math questions, including Logs and complex math.)

http://mathworld.wolfram.com/topics/ComplexNumbers.html

http://www.sciencenewsbooks.org/scibook/imtalstorof1.html (Note: An interesting book on imaginary numbers can be found here.)

Appendix E

COMPUTER RESOURCES

Current spreadsheets have varying degrees of capabilities with complex numbers. In Microsoft Excel, for example, after you install the Analysis Pack (see your documentation) some of the available functions include those shown in Table E-1.

Table E-1 Some Complex Functions Available in Microsoft Excel

	A	B	C
1		Some Excel Complex Algebra Functions	
2			
3	+COMPLEX(50,-100)	50-100i	Expresses a complex number
4	+COMPLEX(25,200)	25+200i	Expresses a complex number
5	+IMREAL(B3)	50	Real part of B3
6	+IMAGINARY(B3)	-100	Imaginary part of B3
7	+IMABS(B3)	111.8033989	This is $|Z|$ = Sqrt(R^2 + X^2)
8	+IMARGUMENT(B3)	-1.10715	Angle, theta, in radians
9	+DEGREES(B8)	-63.43494882	Angle, theta, in degrees
10	+IMSUM(B3+B4)	75+100i	Sum of B3 + B4
11	+IMPRODUCT(B3,B4)	21250+7500i	Product of B3 * B4
12	+IMDIV(B3,B4)	-0.461538461538462-0.307692307692308i	Division of B3 / B4

Mathcad has an outstanding complex math capability. Figure E-1 shows the Mathcad program I used to compare the effects of ESR on two sets of bypass capacitor assumptions. Each set assumes 200 each .01 uF capacitors with 10 nH lead inductance and 100 each 50 uF capacitors with 20 nH inductance. The two curves represent assumptions of .1 Ω and .0001 Ω ESR, respectively. In a more general sense, this is simply an analysis of two parallel RLC circuits.

Appendix E

$C1 := .01 \cdot 10^{-6}$

$R1 := .0001$

$L1 := 10 \cdot 10^{-9}$

Array of 200 $X1(f) := \dfrac{1}{200} \cdot \left(\dfrac{1}{2i \cdot \pi \cdot f \cdot C1} + R1 + 2i \cdot \pi \cdot f \cdot L1 \right)$

$C2 := 50 \cdot 10^{-6}$

$R2 := .0001$

$L2 := 20 \cdot 10^{-9}$

Array of 100 $X2(f) := \dfrac{1}{20} \cdot \left(\dfrac{1}{2i \cdot \pi \cdot f \cdot C2} + R2 + 2i \cdot \pi \cdot f \cdot L2 \right)$

$C3 := .01 \cdot 10^{-6}$

$R3 := .1$

$L3 := 10 \cdot 10^{-9}$

Array of 200 $X3(f) := \dfrac{1}{200} \cdot \left(\dfrac{1}{2i \cdot \pi \cdot f \cdot C3} + R3 + 2i \cdot \pi \cdot f \cdot L3 \right)$

$C4 := 50 \cdot 10^{-6}$

$R4 := .1$

$L4 := 20 \cdot 10^{-9}$

Array of 100 $X4(f) := \dfrac{1}{20} \cdot \left(\dfrac{1}{2i \cdot \pi \cdot f \cdot C4} + R4 + 2i \cdot \pi \cdot f \cdot L4 \right)$

$fstart := 10^5 \qquad fend := 10^9 \qquad N := 5000 \qquad n := 0, 1 .. N \qquad f_n := \left[fstart \cdot \left(\dfrac{fend}{fstart} \right) \right]^{\frac{n}{N}}$

Figure continued next page.

Appendix E

Impedance of first part $\quad Z1_n := |X1(f_n)|$

Impedance of second part $\quad Z2_n := |X2(f_n)|$

Both in parallel

$$Z0001_n := \left| \frac{X1(f_n) \cdot X2(f_n)}{X1(f_n) + X2(f_n)} \right|$$

$$Z1_n := \left| \frac{X3(f_n) \cdot X4(f_n)}{X3(f_n) + X4(f_n)} \right|$$

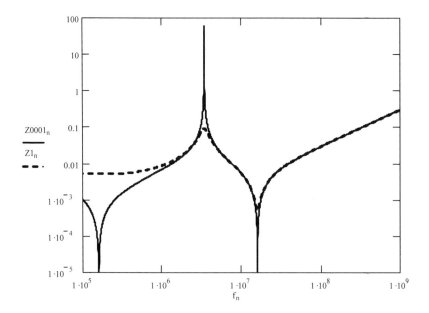

Figure E-1 Mathcad program used to evaluate parallel RLC circuits.

Appendix E

EULER'S FORMULA

A man named Leonhard Euler discovered a mathematical identity related to complex numbers way back in 1748. Think about the resources he had available to him when he did this. His identity is:

$$e^{jx} = \cos(x) + j\sin(x)$$

(In this relationship, x is measured in radians.) Now do two things. Substitute ωt for x and recognize that sin(x) is the same as cos(x − 90°). Then we have:

$$e^{j\omega t} = \cos(\omega t) + j\cos(\omega t - 90°)$$

This looks awfully close to some of the formulas we worked with for impedance and time constants. On the left side we have an exponential, and on the right side we have a real number and an imaginary number shifted 90 degrees in phase. The real relationships that tied these together are too complex for this book, but this gives a hint about how that relationship is going to look.

Here is another interesting thing about Euler's Formula. Suppose you substitute π for x in the first formula. Then, recognize that cos(π) = −1 and sin(π) = 0. Therefore:

$$e^{j\pi} = -1 \text{, or}$$
$$e^{j\pi} - 1 = 0$$

This second form connects the five most common constants in mathematics: e, π, j, 0, and 1.

One more interesting result of this occurs if you substitute π/2 for x in the top equation, and then recognize that cos(π/2) = 0 and sin(π/2) = 1:

$$e^{j\pi/2} = j$$

Now raise both sides of this relationship to the power of j (and note that $j^2 = -1$).

$$e^{-\pi/2} = j^j$$

Using a simple calculator, you can evaluate $e^{-\pi/2}$ and show that

$$j^j = .2078795763......$$

That is, raising an imaginary number to an imaginary power can lead to a real-number answer.

COMPLEX NUMBERS (REPRINTED BY PERMISSION FROM MICROSOFT ENCARTA 2000)

I Introduction

Complex Number, in mathematics, the sum of a real number and an imaginary number. An imaginary number is a multiple of i, where i is the square root of –1. Complex numbers can be expressed in the form a + bi, where a and b are real numbers. They have the algebraic structure of a field in mathematics. In engineering and physics, complex numbers are used extensively to describe electric circuits and electromagnetic waves. The number i appears explicitly in the Schrödinger wave equation, which is fundamental to the quantum theory of the atom. Complex analysis, which combines complex numbers with ideas from calculus, has been widely applied to subjects as different as the theory of numbers and the design of airplane wings.

II History

Historically, complex numbers arose in the search for solutions to equations such as $x^2 = -1$. Because there is no real number x for which the square is –1, early mathematicians believed this equation had no solution. However, by the middle of the 16th century, Italian mathematician Gerolamo Cardano and his contemporaries were experimenting with solutions to equations that involved the square roots of negative numbers. Cardano suggested that the real number 40 could be expressed as

$$(5+\sqrt{-15})(5-\sqrt{-15})$$

Swiss mathematician Leonhard Euler introduced the modern symbol i for $\sqrt{-1}$ in 1777 and expressed the famous relationship $e^{\pi i} = -1$ which connects four of the fundamental numbers of mathematics. For his doctoral dissertation in 1799, German mathematician Carl Friedrich Gauss proved the fundamental theorem of algebra, which states that every polynomial with complex coefficients has a complex root. The study of complex functions was continued by French mathematician Augustin Louis Cauchy, who in 1825 generalized the real definite integral of calculus to functions of a complex variable.

III Properties

For a complex number a + bi, a is called the real part and b is called the imaginary part. Thus, the complex number –2 + 3i has the real part –2 and the imaginary part 3. Addition of complex numbers is performed by adding the real and imaginary parts separately. To add 1 + 4i and 2 – 2i, for example, add the real parts 1 and 2 and then the imaginary parts 4 and -2 to obtain the complex number 3 + 2i. The general rule for addition is

$$(a + bi) + (c + di) = (a + c) + (b + d)i$$

Multiplication of complex numbers is based on the premise that ii = -1 and the assumption that multiplication distributes over addition. This gives the rule

$$(a + bi)(c + di) = (ac - bd) + (ad + bc)i$$

For example,

$$(1 + 4i)(2 - 2i) = 10 + 6i$$

If z = a + bi is any complex number, then, by definition, the complex conjugate of z is

$$\bar{z} = a - bi$$

and the absolute value, or modulus, of z is

$$|z|=\sqrt{a^2+b^2}$$

For example, the complex conjugate of 1 + 4i is 1 − 4i, and the modulus of 1 + 4i is

$$\sqrt{1^2+4^2}=\sqrt{17}$$

A basic relationship connecting absolute value and complex conjugate is

$$z\bar{z}=|z|^2$$

IV The Complex Plane

In the same way that real numbers can be thought of as points on a line, complex numbers can be thought of as points in a plane. The number a + bi is identified with the point in the plane with x coordinate a and y coordinate b. The points 1 + 4i and 2 − 2i are plotted in Figure E-1 and correspond to the points (1,4) and (2, −2). In 1806 Swiss bookkeeper Jean Robert Argand was one of the first people to express complex numbers geometrically as points in the plane. For this reason, Figure E-1 is sometimes referred to as an Argand diagram. If a complex number in the plane is thought of as a vector joining the origin to that point, then addition of complex numbers corresponds to standard vector addition. Figure E-2 shows the complex number 3 + 2i obtained by adding the vectors 1 + 4i and 2 − 2i.

Since points in the plane can be written in terms of the polar coordinates r and Θ, every complex number z can be written in the form

$$Z = r\,[\cos(\Theta) + i\sin(\Theta)]$$

Here, r is the modulus, or distance to the origin, and Θ is the argument of z, or the angle that z makes with the x axis. If

Appendix E

$$z = r[\cos(\Theta) + i\sin(\Theta)] \text{ and}$$
$$w = s[\cos(\Phi) + i\sin(\Phi)]$$

are two complex numbers in polar form, then their product in polar form is given by

$$zw = rs[\cos(\Theta + \Phi) + i\sin(\Theta + \Phi)]$$

This has a simple geometric interpretation that is illustrated in Figure E-3.

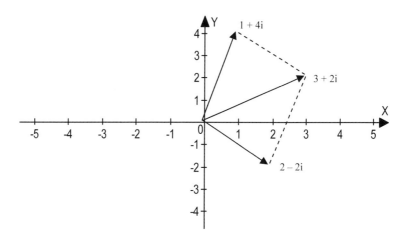

Figure E-2 This graph illustrates the addition of two complex numbers by using vectors in the complex plane with cartesian coordinates. The parallelogram shows that the sum of $1 + 4i$ and $2 - 2i$ is $3 + 2i$.

V Solutions to Polynomials

There are many polynomial equations that have no real solutions, such as

$$x^2 + 1 = 0$$

However, if x is allowed to be complex, the equation has the solutions $x = \pm i$, where i and $-i$ are roots of the polynomial $x^2 + 1$. The equation

$$x^2 - 2x + 2 = 0$$

has the solutions x = 1 ± i. In his fundamental theory of algebra, Gauss showed that every nontrivial (having at least one nonzero root) polynomial with complex coefficients must have at least one complex root. From this it follows that every complex polynomial of degree n must have exactly n roots, although some roots may be the same. Consequently, every complex polynomial of degree n can be written as a product of exactly n linear, or first-degree, factors.

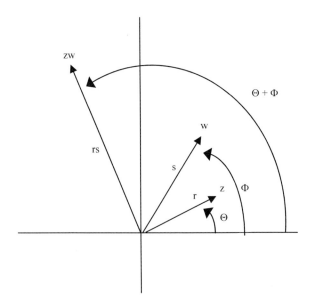

Figure E-3 This graph illustrates the multiplication of two complex numbers by using vectors in the complex plane with polar coordinates. The product of vectors z and w is a vector whose length is the product of the lengths of z and w, and whose angle with the x axis is the sum of the angles that z and w make with the x axis.

Reprinted from: "Complex Number," Microsoft Encarta Online Encyclopedia 2000, *http://encarta.msn.com*. Copyright 1997–2000 Microsoft Corporation. All rights reserved.

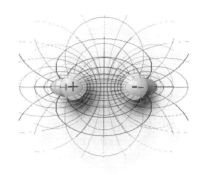

F TRANSMISSION LINE SIMULATOR

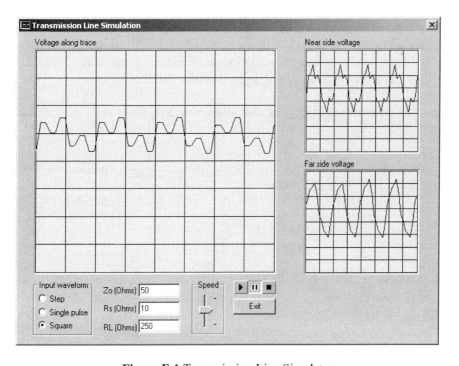

Figure F-1 Transmission Line Simulator

*T*his simulator can be downloaded from
http://www.ultracad.com/prenticehall/tlinesim.zip.

Unzip the file to obtain tlinesim.exe and execute the file. The program shown in Figure F-1 will start.

The main screen represents a transmission line. The pattern represents the signal at every point along the line in real time. The left edge of the screen at the upper right represents the instantaneous signal level at the very beginning of the trace, and the left edge of the screen at the lower right represents the same thing for the far end of the trace. In this way, you can visualize what is happening at every point along the line.

The Input waveform selection enables you to select a single step-function change in voltage, a single pulse, or a repetitive square wave (much like a clock signal) as the driving signal. The three text boxes allow you to specify the characteristic impedance of the transmission line, and the resistive loading at each end. Having set these, the right arrow then starts the selected waveform down the trace. The waveform will reflect back and forth for at least 5 cycles, illustrating what happens to the waveform under various termination conditions. The speed at which the waveform travels is controlled by the speed selector slider.

The other two selection buttons allow you to pause or reset the simulation.

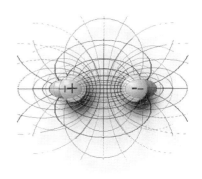

G   ECHO ILLUSTRATION

A clear message (signal) will be distorted if there are any reflections from the end of the transmission line. One solution to the reflection problem is to slow down the message (slow down the rise time of the signal). UltraCAD has placed three audio files on its Web site that illustrate these points. The first, testwave1.wav, is a simple phrase. The second, testwave2.wav, is the same phrase with considerable (admittedly forced) echoes (reflections) added. The third, testwave3.wav, is the same expression with the same echoes, but the phrase is spoken more slowly.

The three files illustrate how reflections can obscure a message, and how slowing down the message can help improve the receiver's ability to understand it. The three files can be found at:

- *http://www.ultracad.com/prenticehall/testwave1.wav*   (76k)
- *http://www.ultracad.com/prenticehall/testwave2.wav*   (76k)
- *http://www.ultracad.com/prenticehall/testwave3.wav*   (376k)

Your browser may play the file directly. Alternatively you might get better results if you go to *http://www.ultracad.com/prenticehall,* right-click each file individually, and download it to your computer. Then you will be able to play the file locally. Of course you will not be able to play the file if you do not have an audio card and speakers.

This illustration is admittedly exaggerated. It is not intended to accurately reflect what happens on a transmission line. Its purpose is to suggest, by analogy, what happens when transmission lines are not terminated properly and reflections are allowed to exist on a line.

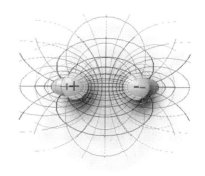

H UltraCAD FREEWARE CALCULATORS

*U*ltraCAD Design, Inc. has developed several freeware calculators for people to use. They are all located on the UltraCAD Web site at *www.ultracad.com/calcs*.

This appendix describes three of them that are useful for signal integrity purposes. As a general warning, be sure to read the help file associated with each calculator before using it.

IMPEDANCE (ULTRACLC.EXE)

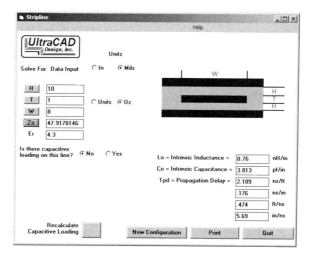

Figure H-1 A typical stripline screen from UltraCAD's impedance calculator.

This tool, shown in Figure H-1, can be used to determine the trace width, thickness, or stackup dimensions required to hit an impedance target, or it calculates the trace impedance given the other parameters. It gives helpful results for microstrip, embedded microstrip, stripline, and asymmetric stripline configurations. The calculations are based on the standard equations found in the industry and summarized at the end of Chapter 10.

Note: This calculator is not designed to be used for differential impedance calculations.

CROSSTALK (ULTRA_CT.EXE)

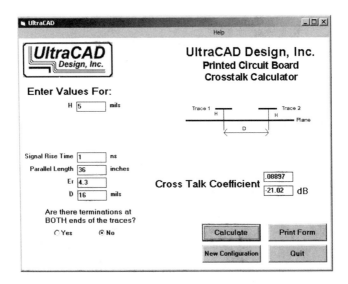

Figure H-2 A typical screen from UltraCAD's crosstalk calculator.

UltraCAD's crosstalk calculator, shown in Figure H-2, is useful for making estimates of the crosstalk coefficient between adjacent traces. Since there are many factors that can affect crosstalk, the calculator is not intended to give precise results. Rather it gives an approximate, worst-case indication. People should look at existing boards with this calculator and calibrate the calculator to previous designs.

Appendix H

BYPASS CAP AND ESR (UCADESR3.EXE)

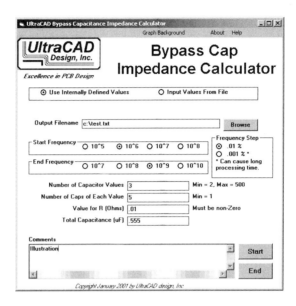

Figure H-3 One of the screens from UltraCAD's Bypass Cap Impedance Calculator.

The equations for a system of parallel capacitors (each with lead inductance and ESR) get very complicated very quickly. The manual calculation of results becomes impractical when the number of parallel capacitors reaches three or four. Unfortunately there are very few, if any, practical alternatives for making these calculations.

UltraCAD designed this calculator (see Figure H-3) for its own internal use for making these types of calculations. It can determine the resonant frequencies and plot an impedance curve for a system of up to 500 different capacitor values with an effectively unlimited number of capacitors of each value. A typical graphical output is shown in Figure H-4. Output data is also written to a text file. The useful frequency range of analysis is from 100 kHz to 10 GHz.

The calculator operates in two modes. The first is a demonstration mode in which the calculator defines the capacitance and inductance values over a particular range, based on some general user inputs. The second mode accepts virtually any user-specified values through an easily programmed input file.

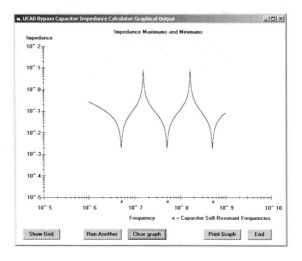

Figure H-4 Representative output from UltraCAD's bypass capacitance calculator.

The freeware version of this calculator accepts up to three capacitance values. A nominal license is necessary to extend the capability of the calculator beyond that limitation.

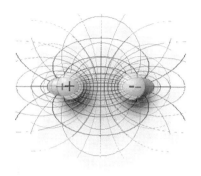

I TDRs and VNAs

*I*t is not uncommon now to read an article or see a presentation with references to a time domain reflector (TDR) analysis or to see a graph of an analysis based on a TDR response. Less commonly we are beginning to see references to similar analyses with vector network analyzers (VNAs). Many designers have never had the opportunity to learn what these tools do or how they work. This appendix gives an overview of the operating principles behind the tools so you can at least understand what it is they are able to show.

BASIC CONCEPTS

In one sense, the operating principles behind TDRs and VNAs are almost identical. They have very similar block diagrams, analyze the same circuits, and give equivalent results. Both are used to analyze impedance-controlled traces or networks. Figure I-1 illustrates the block diagram that applies to each.

Each has a signal-generating circuit we can model as a Thevenin equivalent circuit (see Chapter 5). This is shown in the left section of Figure I-1. The series (output) resistor is shown as 50 Ω, which is typical, but in some instruments the output impedance can be varied by the user. Each has a signal analysis circuit at the output of the tool to analyze the signal and its return from the device under test. Provision must be made for some sort of output, but this can take many different forms. The output might be a built-in display, it might be an output to a plotter, or it might be a text output to a screen or a file. In some cases (see Figure I-2) it might be an output to a separate computer.

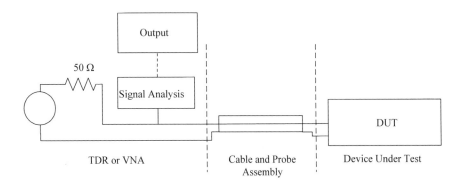

Figure I-1 Block diagram of a TDR or VNA test setup.

The device under test (DUT) is shown as the right section in the figure. This can be an actual circuit board or it can be a specially designed or processed test board. It can also be a real-world network of some type. For example, TDRs can be used to determine where a fault is along a telephone line between the central office and your home.

Each product must be connected to the DUT by a cable and probe assembly, shown as the middle section. This would normally be an impedance-controlled cable that would match the source impedance of the tool. If the TDR or VNA is used to analyze the impedance characteristics of a DUT, it is clear that the cable and probe assembly should match the impedance of the testing tool, so as not to introduce any erroneous indications into the analysis. Attaching the cable and probe assembly to the DUT in such a way as to introduce no additional impedance discontinuities can be *the* most difficult part of any TDR or VNA analysis.

Now consider what happens when the tool (TDR or VNA) sends a signal through the cable to the DUT. Let's assume the cable and probe assembly is 2 ns long and the trace under test is 3 ns long. The signal analysis circuit looks at the signal at the output of the tool. At first it sees only the output of the tool. It takes 2 ns for this output to travel to the front of the DUT. If there is any kind of reflection from the front of the DUT (e.g., the probe connection is not exactly 50 Ω) there will be a reflection from that point back to the tool. This reflection will arrive 2 ns later at the signal analysis circuit. Any additional impedance discontinuities will also cause reflections that will travel back to the signal analysis circuit. By measuring the signal level and its time delay, and comparing that to the source signal, the signal analysis circuit has all the information it needs to characterize the DUT.

Appendix I

TDR vs. VNA

TDRs and VNAs have one very significant difference between them. The TDR transmits a step-function change in signal to the DUT. For simplicity, let's think of that signal being a *very* fast transition from a logical zero to a logical one, or from 0.0 volts to 1.0 volt. Rise times in the picoseconds are not uncommon. This signal travels down the DUT and returns. By analyzing the signal level and the time it took for the signal to travel over its distance, any *specific, individual point* along the trace can be characterized. Because the TDR looks at the response as a function of time, it is referred to as a *time domain* device, hence the name. It looks at reflections as a function of time.

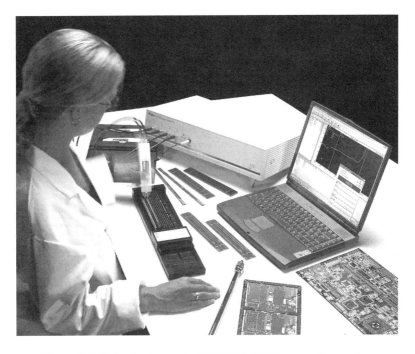

Figure I-2 Polar Instruments CITS500s TDR measuring system using a standard PC for data analysis, processing, and display (courtesy Polar Instruments).

By contrast, a VNA sends a longer (sine-wave, AC) signal down the trace at a specific frequency. The pulse width of this AC signal is long enough for the reflections from the DUT to stabilize, but it is still a very short period of time in real terms. The signal analysis circuit then measures the stabilized response, calculates the impedance of the DUT, and calculates the phase shift of the returned signal. Then the VNA shifts to another frequency and does this analysis all over again. After a period of time, the VNA has "swept" over a range of frequencies and characterized the DUT at each one of them. The output of the test is a plot of the impedance and phase shift as a function of frequency. Since the VNA looks at the response of the DUT as a function of frequency, it is referred to as a *frequency domain* device.

Thus, a TDR is a time domain tool that looks at the impedance characteristics of a trace at each individual point along the trace. A VNA is a frequency domain device that looks at the average impedance and phase shift of an entire trace at each individual frequency across a range of frequencies.

PRINCIPLES OF OPERATION: TDR

Figure I-3 is a TDR response taken directly from a study done in the early 1990s.[1] It illustrates a trace exactly 15 inches long with three equally spaced vias along it. The trace passes through each via to another signal layer on the test board. All traces are designed to be 50-Ω traces. There is a high-frequency 50-Ω SMA connector at the front of the trace.

As is typical of TDR analyses, the graph does not show the portion of the signal travel along the cable and probe assembly. Thus, it starts at 20 ns after the pulse from the TDR. At about 21.4 ns there is a sharp, momentary drop in the signal level. This is the point at which the signal travels through the connector, which (from this figure) is obviously not exactly 50 Ω. Then the signal level rises above the reference, indicating that the actual impedance of the trace is higher than the targeted 50 Ω.

The vias are spaced at intervals of 3.75 inches. Each via shows up as a momentary drop in signal level, meaning that each one looks somewhat capacitive to the circuit. There is a 50-Ω terminating resistor at the end of the trace, beyond which the graphical output signal is relatively flat.

---

[1] Brooks, Douglas, "The Effects of VIA on PCB Traces," *Printed Circuit Design*, August 1996, available at *www.ultracad.com*.

Appendix I

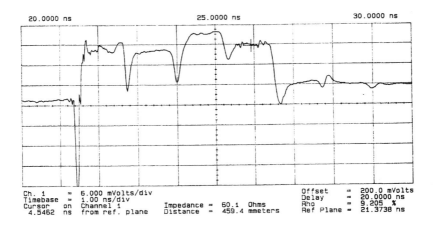

Figure I-3 TDR output showing a 15-inch trace with three equally spaced vias.

Note that the time period along the chart from the signal drop at the connector to the signal stabilization at the terminating resistor is almost exactly 5 ns. This represents the round trip time for the signal to travel down the 15-inch trace and back again. That means the one-way trip is 2.5 ns, which is almost exactly 6 inches/ns, a comforting result.

At each impedance discontinuity there is a reflection. If we call the signal the *incident* wave, then there will be a portion of that wave that is reflected backward and a portion that is transmitted forward. If our TDR is designed to have a Zo of 50 Ωs, and the impedance at the discontinuity is Ztest, then the relationships shown in Equations I-1 and I-2 hold:

$$\frac{V_{reflected}}{V_{incident}} = \rho = \frac{Z_{test} - 50\Omega}{Z_{test} + 50\Omega} \qquad [\text{I-1}]$$

so

$$Z_{test} = \frac{1+\rho}{1-\rho}(50) \ \Omega \qquad [\text{I-2}]$$

Thus, the signal analysis circuit can calculate the specific impedance, Ztest, at each individual point along the trace. The waveform pictured in Figure I-3 is just such a plot.

You might pick up on one real potential problem, though. Consider the reflection at, for example, the second via. There will be a reflected waveform from that via flowing back toward the TDR. But this reflection will flow past the first via, from which there will also be a reflection, and also past the connector, from which there will be an additional reflection. Each of these reflections will have their own incident, reflected, and transmitted components. How, then, do we separate out all these secondary reflections from the primary ones we are interested in? The straightforward answer is that it isn't easy.

The other answer is that it is done through software. It is common in the industry that such software is provided by third-party vendors rather than the TDR hardware manufacturers. Thus a well-equipped analytical test bench will commonly have one or more TDRs from hardware suppliers and also some analytical software from other suppliers.

PRINCIPLES OF OPERATION: VNA

When we send a steady state frequency along the DUT, we also get a transmitted and reflected wave. But since we are dealing with average conditions (instead of instantaneous ones) we use different terms. The terms are called *S parameters*.[2] (Think of S as meaning "scatter," as in the DUT starts *scattering* the waveforms.) In this context, we use the parameters S11 and S12. S11 represents the part of the signal that is going into terminal 1 (the front end of the transmission line under test) and reflected back from the same terminal. S12 represents that part of the waveform that is going on from terminal 1 toward terminal 2 (the other end of the transmission line under test). The analysis is very similar to the TDR case, but now we refer to Ztest as the *average* impedance of the DUT, not the instantaneous impedance at a particular point along the DUT. Equations I-1 and I-2 now become:

---

[2] For an excellent discussion of S parameters, see Bogatin, Eric, "Get Your S (Parameters) Together," *Printed Circuit Design*, February 2003.

Appendix I

$$\frac{V_{reflected}}{V_{incident}} = S11 = \frac{Z_{test} - 50\Omega}{Z_{test} + 50\Omega} \qquad [\text{I-3}]$$

$$Z_{test} = \frac{1+S11}{1-S11}(50)\ \Omega \qquad [\text{I-4}]$$

The S parameter is a complex function that has a magnitude and a phase. For a perfect transmission line, with perfect termination, S11's magnitude and phase would be zero (because there is no reflection). For a uniform 50-Ω transmission line, open at the far end, the magnitude of S11 would be 1.0 (100% reflection). But the phase of S11 would be more complicated.

To visualize what happens from a phase standpoint, assume we send a 1.0-GHz signal down the open line. Assume also that the propagation time is exactly 6 in./ns. The wavelength of the signal would be

$$\frac{1}{(1.0)10^9 \frac{\text{cycles}}{\text{sec}}} \left(\frac{6}{10^{-9}}\right) \frac{\text{in.}}{\text{sec}}$$

$$\left(10^{-9} \frac{\text{sec}}{\text{cycle}}\right)\left(\frac{6}{10^{-9}} \frac{\text{in.}}{\text{sec}}\right) = 6\ \text{in./cycle}$$

Suppose the line were exactly 3 inches long. The round trip down the line and back would be 6 inches. So, what would the phase relationship be between the incident waveform and the reflected waveform? The answer is they would be exactly in phase. The reflection would return exactly one cycle (360 degrees) after it started down the line. We could say that the reflected wave was lagging the incident wave by 360 degrees, or that the incident wave was leading the reflected wave by 360 degrees. But for all practical purposes, that is the same thing as saying that they are actually in phase.

Now, assume we increase the frequency to 1.25 GHz. When the reflected wave returns to the front of the DUT, the incident wave will have progressed through 360 + 90 = 450 degrees. The reflected wave is now lagging the incident wave by an additional 90 degrees. As we increase the frequency slightly, the reflected wave lags further behind. At an incident frequency of 1.5 GHz, the reflected wave will be lagging by an additional 180 degrees. At 1.75 GHz, it will be lagging by an additional 270 degrees. At 2.0 GHz, it will be lagging by another 360 degrees again.

Now here is a (perhaps) philosophical question for you: At 1.75 GHz, is the reflected wave lagging (minus sign) the incident wave by 270 degrees, or is it leading (plus sign) it by 90 degrees? At 1.5 GHz, is the reflected wave lagging (minus sign) the incident wave by 180 degrees or is it leading (plus sign) it by 180 degrees? In each case, both answers are correct. We normally define the phase relationship by a convention that says as we pass 180 degrees the phase shift changes from a lead to a lag, or vice versa. Thus, in this example, the phase relationship of this experiment with frequency is graphed as shown in Figure I-4. This is one place in electronics (passing through 180-degrees phase shift) where we allow an instantaneous (no rise or fall time) change between values.

By looking at both the average impedance of the DUT and the phase shift of the returned signal relative to the VNA signal as a function of the frequency of the VNA test signal, engineers can characterize the full nature of the transmission line under test.

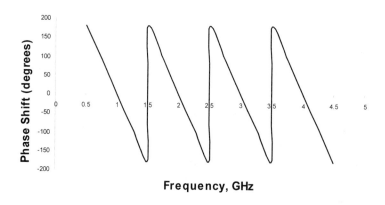

Figure I-4 VNA phase shift from the open end of a 3-inch line.

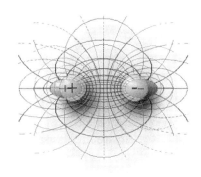

J RIGHT ANGLE CORNERS

*T*he influence of 90-degree corners on neighboring circuits is one of the most controversial topics in PCB design. It ranks right beside the routing of bypass capacitor leads as a guaranteed argument starter.

The arguments against 90-degree corners fall into two categories: impedance control and EMI. I summarize each one of those here.

IMPEDANCE CONTROL

Unless very carefully constructed, a trace rounding a corner will necessarily be wider at the corner than it is in the middle of the trace. Since control over impedance requires a controlled geometry, and since a trace that changes width does not have a controlled geometry by definition, then necessarily there must be an impedance discontinuity (a change in impedance) at that point. The concern, then, is whether this change in discontinuity will cause a reflection down the trace.

Figure J-1 illustrates several geometric trace configurations, the maximum change in width that occurs as the trace rounds the corner, and the approximate impact on the calculated trace impedance. Since the trace gets wider as it goes around the corner, the capacitance between the trace and the plane increases and the impedance goes down. The exact impact on the impedance depends on the magnitude of the other variables. That's why the impact is shown as a range.

The idea of a true right-angle corner is a misnomer. Since traces are typically "drawn" with a laser plotter using a circular spot, the traces on our films really look like (b) in the figure. The traces have a sharp edge at the inner corner but a rounded edge at the outer corner. So, at most the impact on impedance, if we let our design systems route right-angle corners, is 10% or less. But that is at a single, sharp, well-defined point. The impact is less than that everywhere else. Remember the concept of critical length introduced in Chapter 10? For a trace width of 6 mils we are talking in the terahertz range before this distance approaches the critical length. So in reality there is no impedance impact at all for signals rounding corners.

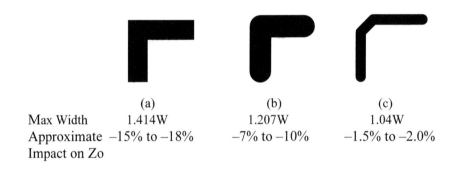

	(a)	(b)	(c)
Max Width	1.414W	1.207W	1.04W
Approximate Impact on Zo	−15% to −18%	−7% to −10%	−1.5% to −2.0%

Figure J-1 The maximum trace width and impact on Zo is a function of trace geometry.

If that answer doesn't convince you, then think of the impedance discontinuity that we experience around vias. A via is a much worse geometrical perturbation and is so over a much longer distance than a corner. We intuitively know that vias generally don't present a serious discontinuity problem, so the impedance question related to 90-degree corners is really a nonissue.

EMI R<small>ADIATION</small>

A different argument against right-angle corners goes like this. There is electromagnetic energy surrounding a high-speed trace, by definition. The fields, and therefore that energy, become concentrated at the corners. The sharper the corner, the greater the concentration of energy. Therefore, right-angle corners lead to radiation problems at these points of concentration.

Appendix J

The impedance argument can be argued theoretically. The radiation argument, on the other hand, does not lend itself to an easy theoretical analysis. The argument sounds good, but the proof is really in the empirical analysis.

UltraCAD participated in a study in which a test board was designed and fabricated with several different kinds of corners for test.[1] These corners ranged from smoothly curved arcs to radically diabolical 135-degree sharp bends (see Figure J-2). These traces were then tested for radiation in a controlled test. There were no significant differences in the radiation levels from any of the traces at frequencies up to about one GHz (the upper limit of the test).

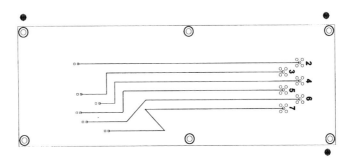

Figure J-2 Partial layout of an EMI test board.

CONCLUSION

On circuit boards, at least up into the gigahertz range, there is no particular performance reason to avoid 90-degree corners. We at UltraCAD avoid them anyway. We think mitered corners look better, but we cannot argue a performance reason for taking the time for the mitering process.

---

[1] "90 Degree Corners, The Final Turn," *Printed Circuit Design*, January 1998, copy available at *www.ultracad.com*.

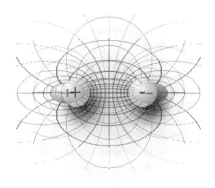

GLOSSARY

*N*ote: Many of these entries are explained in nontechnical terms for ease of understanding. They are explained principally as they relate to PCB design issues. *sy* = symbol; *typ* = typical usage in our industry; *val* = value of a constant or formula relationship.

AC (alternating current) A current that repeatedly changes value or changes direction with time.

Admittance *sy* Y, *val* 1/Z, the inverse of impedance.

Anti-resonance *typ* Term used to describe the parallel resonance that can occur between two capacitors wired in parallel.

Atom A basic physical building block consisting of protons, neutrons, and electrons.

Backward crosstalk Coupled noise from an aggressor trace to a victim trace that travels in a direction opposite to that of the aggressor signal.

Bandwidth *typ* A range of frequencies that a circuit passes or handles with minimal distortion.

Capacitive coupling   A signal that couples from one circuit, wire, or trace to another circuit, wire, or trace caused by changing levels of electrical charge.

Capacitor, capacitance   A device or condition that opposes the flow of current while causing a –90-degree voltage phase shift. Also a device that stores charge.

Capacitor, bypass   *typ* A capacitor with the primary function of stabilizing a power supply voltage, especially for an adjacent device or circuit.

Charge   A measure of electrical energy. Particles with like charge repel each other. Particles with opposite charge attract each other. The forces are proportional to the magnitude of the charge and inversely proportional to the square of the distance.

Charge, negative   A negative unit of charge. Electrons have negative charge. Charge is often relative; that is, one point is negatively charged relative to another point, not necessarily in absolute terms.

Charge, positive   A positive unit of charge. A proton has positive charge. Positive charge often results from a lack of electrons (negative charges). Charge is often relative; that is one point is positively charged relative to another point, but not necessarily in absolute terms.

Common log   *sy* Log   Logarithm taken to the base 10, (cf. *Natural log*).

Common mode current   *typ* Refers to a return signal when the current flows where you do not expect it to flow.

Common mode signal   *typ* Signals that are equal in magnitude and phase at the inputs of a device.

Complex number   *typ* A number containing both a real and an imaginary part.

Conductance   *sy* G, *val* = 1/R, the inverse of resistance.

Constant current generator A source that generates, or supplies, a constant current regardless of load. This would imply a very high voltage if the load is very high impedance.

Constant voltage generator A source that generates, or supplies, a constant voltage regardless of load. This implies it can supply a very large amount of current if the load impedance is very low.

Coulomb An amount of charge equal to 6.25×10^{-18} electrons.

Critical length A trace length for which the propagation time equals one-half the signal rise time.

Crosstalk A signal induced in one wire or trace by current in another wire or trace.

Current *sy* i, Flow of electrons.

DC (direct current) A current that flows in one direction, usually without changing magnitude.

di/dt A change of current divided by a change in time for a very short time period. Often, dt represents the rise or fall time.

Dielectric *typ* The insulating material between trace layers.

Dielectric absorption A condition in which the molecules in the dielectric begin to absorb very high-frequency energy causing signal losses at the higher-frequency harmonics.

Dielectric constant A measure of an insulator's ability to store charge. It actually may not be a constant under certain conditions.

Dielectric constant, relative $sy = \varepsilon_r$, Dielectric constant of a material divided by the dielectric constant of a vacuum.

Differential impedance The impedance seen by a driver when a pair of traces are driven differentially.

Differential mode   *typ* Refers to a signal and its return when the current flows where you expect it to flow.

Differential signals   *typ* A signal paired and routed in close relationship with its return, often for improved signal–noise ratio.

Differential traces   A pair of traces routed in relationship to each other for conducting differential signals.

DUT   Device under test.

e   A constant = 2.718282....., the base of the natural logarithm.

Electromagnetic coupling   A signal induced in another conductor as a result of electromagnetic radiation.

Electromagnetic radiation   A current radiates both an electrical field and a magnetic field. Collective name for these fields.

Electron   An atomic particle (circling around the nucleus) containing negative charge. There are as many electrons in an atom as there are protons.

EMC (electromagnetic compliance)   Compliance to rules and regulations controlling EMI.

EMI (electromagnetic interference)   Interference caused by electromagnetic radiation.

Equalization   *typ* A technique for compensating for frequency-related losses along transmission lines.

Equivalent circuit   A simplified circuit with outputs that are the same as those of a more complex circuit. Typically contains only a power source and an impedance.

ESR (equivalent series resistance)   *typ* The real part of the impedance expression for a capacitor.

Exponent The power a number is raised to. n is an exponent in the expression a^n.

Exponential *typ* Any function containing the constant e raised to a power.

Eye diagram Method for visualizing a large number of changing logic states at a point in a system under a large number of signal conditions.

Fall time Time required for a signal to fall from its highest value to its lowest value. Typically measured between the 90% and 10% points on the curve, but sometimes between the 80% and 20% points.

Farad A measure of capacitance.

Forward crosstalk Coupled noise from an aggressor trace to a victim trace that travels in the same direction as the aggressor signal.

Fourier analysis An analytical technique for separating a complex mathematical waveform into a function or summation of a number of simpler trigonometric waveforms.

Frequency, angular *sy* ω, *typ* A frequency measure based on radians. $\omega = 2\pi F$.

Frequency, definition *sy* F, *typ* The number of cycles a sine wave completes in one second.

Frequency domain Analyses performed as a function of frequency. When graphed, frequency would typically be on the horizontal axis, (cf. *time domain*).

Generator Device that converts mechanical motion to electrical energy. Can be the same device as a motor.

Giga 1,000,000,000 or 10^9.

Guard band A grounded trace typically placed between two signal traces for the purpose of controlling crosstalk.

Harmonic   *typ* Frequencies that are multiples of a fundamental frequency.

Henry   A measure of inductance.

Imaginary operator   *sy* j, *val* the square root of –1. Since there is no number which, when multiplied by itself, can be negative, it is called imaginary.

Imaginary part   *typ* Any expression for voltage, current, or impedance that is multiplied or divided by j and is shifted ± 90 degrees in phase from the real part of the same expression.

Impedance   *sy* Z, *val* R + jX, Opposition to the flow of current, typically with both resistive and reactive components but may contain only either one.

Impedance, characteristic   *sy* Zo, *typ* The input impedance of a transmission line.

Impedance, common mode   The impedance seen by a driver when a pair of traces is driven with a common signal.

Impedance, differential   The impedance seen by a driver when a pair of traces is driven differentially.

Impedance, single-ended   *sy* Zo or Z11, The characteristic impedance of an individual trace, even if it is part of a differential pair.

Impedance, transmission line   *sy* Zo, *typ* The input impedance of a transmission line.

Inductive coupling   A signal that couples from one circuit, wire, or trace to another circuit, wire, or trace caused by a changing magnetic field.

Inductor, inductance   A device or condition that opposes the flow of current while causing a +90-degree voltage phase shift.

j Constant, the square root of –1. The symbol i is used in every other discipline, but is not used in electronics to avoid confusion with the symbol for current.

Kilo 1,000, or 10^3.

Logarithm A numeric operation. If a base B is raised to the a power, then the log, to the base B, of that operation is a. $\text{Log}_b(b^a) = a$. See also *common* or *natural log*.

Loop area An area defined by the signal path and its return path, relevant for EMI considerations.

Mega 1,000,000, or 10^6.

Micro .000001, or 10^{-6}.

Microstrip A trace configuration where there is a reference plane on only one side of a signal trace.

Milli .001, or 10^{-3}.

Motor A device that converts electrical energy to mechanical motion. Can be the same device as a generator.

Nano 10^{-9}.

Natural log *sy* Ln, Logarithm taken to the base e.

Neutron An atomic particle (part of the nucleus) containing no charge.

Ohm *sy* Ω, A measure of resistance.

Ohm's Law Relationship between voltage, current, and resistance. *typ* $E = I \times R$.

Period *sy* τ, *val* 1/F, The time required for one complete waveform cycle. The inverse of the cyclical frequency.

Phase Relationship of the angular position on a sinusoidal waveform compared to a reference, such as another waveform.

Phase shift *typ* The difference in the phase between a voltage and a current waveform, or a difference caused by some other factor, such as impedance.

Pi *sy* π, A constant = 3.14159.......

Pico 10^{-12}.

Propagation speed The speed with which a signal propagates along a trace. Often expressed in units of distance per unit time. The inverse of propagation time.

Propagation time The time required for a signal to move (propagate) along a trace.

Proton An atomic particle (part of the nucleus) containing positive charge. There are as many protons in an atom as there are electrons.

R *typ* Resistance or the real part of a complex number.

Radian The angle formed by a length along the circumference of a circle equal to the radius. There are 2π radians in 360 degrees.

Reactance *sy* X, Resistance to a flow of current with a 90-degree phase shift.

Real part Any numeric expression for voltage, current, or impedance that is not multiplied by j and does not have a phase shift.

Reflection coefficient *sy* ρ, *val* (Zo – RL)/(Zo + RL), its value ranges from +1 to –1. It determines the magnitude of the reflection at the end of a transmission line.

Relative dielectric constant *sy* ε_r, Dielectric constant of a material divided by the dielectric constant of a vacuum.

Resistance   *sy* R, Resistance to the flow of current without causing a phase shift.

Resistor   A device that opposes the flow of current without causing a phase shift.

Resonance   A condition in which the reactance term goes to zero. *typ* Where the impedance function peaks or minimizes.

Resonant frequency   The frequency where the reactance term goes to zero.

Right-hand rule   A rule that helps determine the relationship between the direction of current flow and its accompanying magnetic field.

Rise time   Time required for a signal to rise from its lowest value to its highest value. Typically measured between the 10% and 90% points on the curve but sometimes between the 20% and 80% points.

RMS   Root Mean Square. For an AC waveform it is the equivalent DC value that would generate the same power.

Signal-to-noise ratio   The ratio of a circuit signal to the electrical noise in the circuit. In general, higher ratios are better.

Skin effect   Tendency for the current density to be greater at the edge of a wire or trace instead of being evenly distributed over the entire cross-section. Effect increases with frequency.

Step function   *typ* Suggests an instantaneous change from one voltage level to another or one logic state to another. Since instantaneous changes are not possible, step function implies a very rapid rise or fall time.

Stripline   A circuit board configuration in which a signal trace is placed between two reference planes.

Stub   *typ* A short trace extending from a main trace to a component pin.

Susceptance   *sy* B, *val* 1/X, the inverse of reactance.

TDR   Time domain reflectometer; an instrument for measuring impedance.

Termination   One or more components used in conjunction with a transmission line to control signal reflections.

Time constant   A measure calculated as 1/RC or L/R. Any RC or RL circuit typically exhibits very predictable behavior when looked at in terms of time constants.

Time domain   Analyses performed as a function of time. When graphed, time would typically be on the horizontal axis, (cf. *frequency domain*).

Transmission line   *typ* A tightly geometrically controlled trace structure, typically with respect to a reference plane or other return path.

Tuned trace   A trace trimmed to a precise length to achieve a timing objective.

VNA   Vector network analyzer; an instrument for measuring impedance and circuit characteristics.

Voltage   An electrical force caused by the difference in charge between two points.

Wavelength   *sy* λ, The distance a signal travels in one complete cycle, (cf. *propagation time*).

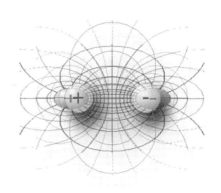

INDEX

2
20-H rule 173

A
AC (Alternating Current) 8, 10, 12, 49, 74, 157, 164, 305, 340, 378, 387
Admittance 142, 387
Aggressor ...221, 223, 238, 242, 387, 391
Ampere, André 5, 26, 155, 329
Amplifier gain 85
Antenna 36, 91, 92, 156, 168, 180
Anti-resonance...........................292, 387
Arctangent 124
Atom 3, 387
Atomic number......................................3
Atomic weight 3

B
Bandwidth 21, 137, 387
Broadside coupled 268

C
Capacitive coupling.....................222, 257
Capacitive loading......................188, 206
Capacitive reactance 98
Capacitor, capacitance 388
       bypass...273

calculating 48, 282
hydraulic analog 43
impedance.................................. 278
parallel... 45
phase shift........ *See* Phase angle, shift
planar...................275, 282, 295, 300
RC time constant 74, 132
series.. 45
time constant............................. 194
voltage and current through 64
Characteristic impedance........................
       See Impedance, transmission line
capacitive loading....................... 206
Charge 27, 40, 155, 296, 329, 388
    AC termination 194
    bypass capacitor 273
    capacitor 43, 47, 57, 64
    conservation of 78
    coulomb....................................... 5
    crosstalk................................... 221
    direction of flow 23
    farad.. 47
    relative dielectric coefficient 28
Common mode 169, 249, 255, 388
Communications model...................... 175
Complex number 388
Conductance 94
Coulomb 5, 64, 389
Coulomb, Charles 1, 5, 26, 329

Index

Coulomb's Law 1, 6, 329
Coupling
    crosstalk 221
    differential 257, 261
    EMI 159
    power system 304
    to return signal 343
Critical length 183, 389
Crosstalk 300, 340, 389
    backward 223, 387
    estimating 226
    forward 225, 391
    simulation 235, 239, 242, 244
Current 3, 4, 26, 56, 221, 389
    and Ohm's Law 39
    and reactance 97
    changing 155, 156, 159, 222, 312, 314, 334, 338, 340
    changing through inductor 54
    charge/discharge 56
    charge/discharge in AC termination 194
    charging a capacitor 64
    constant within loop 152, 251
    coupling 159
    coupling to adjacent trace 343
    coupling to plane 257
    crosstalk coupling 221
    definition 5, 25
    density 302, 313
    density under trace 228, 298
    direction of flow 23
    EMI coupling 156
    flow of electrons ... 1, 3, 6, 23, 38, 64, 78, 152, 155, 158, 221, 273, 299, 332, 389
    flows in closed loop 152, 158, 251, 257, 271
    ground loop 303
    ideal source 121
    loop 161, 165, 193, 300
    loop area in differential pair 256
    return .. 157, 159, 160, 165, 193, 197, 251, 300, 305
    return from trace layers 198
    return from unmatched pairs 253
    return on plane 343
    skin effect 312
    transformer 338

D

DC (Direct Current) 8, 12, 13, 49, 54, 160, 164, 192, 212, 222, 277, 303, 389
di/dt 66, 67, 271, 335, 389
Dielectric absorption 314, 389
Differential impedance 260, 389
    broadside coupled 268
    calculating 266
    edge coupled 266
    simulation 262
Differential mode 169, 170, 249, 390
Differential traces 390
    advantages 249
    design rules 255
    EMI 252, 257
    timing 32
Downware sloping curve 280
dV/dt 64, 298

E

e 73, 346, 390
Edge coupled 266
Electromagnet 155
Electromagnetic coupling .. 159, 221, 343, 390
Electromagnetic field 27, 155
Electromagnetic interference .... See EMI
Electron 3, 27, 43, 155, 221, 269, 273, 387, 388, 390
    flow .... See Current
    shell 3
EMI 156, 202, 255, 390
    20-H rule 173
    changing trace layer 165, 199
    common mode 169
    differential pair loop 257
    differential trace length 252, 253
    Faraday shield 174
    impedance peaks 284

loop area 161, 167
pad clearance 164
picket fence 174
plane pairs 300
reference plane 300
return path 163
right-angle corner 384
slots in planes 162
stripline environment 168
stubs 168
Thevenin termination 193
twisted pair 157
unrelated plane 167
Equalization.......................................390
active 319
passive 319
Equivalent circuit 79, 89, 215, 390
Norton 89
Thevenin....................... 89, 211, 375
ESR (equivalent series resistance)..... 107, 137, 285, 296, 358, 373, 390
and anti-resonance..................... 292
and anti-resonant peak................ 286
and self-resonance..................... 292
and system resonance 291
Even mode.............................. 249, 256, *See also* Common mode
Eye diagram........................... 316, 391

F

Fall time.............................. *See* Rise time
Farad 25, 47, 391
Faraday shield...................................174
Faraday, Michael 26, 47, 156, 329
Faraday's Law 156, 329
Fourier 18, 391
impulse signal............................... 22
sawtooth wave 22
triangular wave 22
Frequency 76, 97, 121, 128
angular 18, 98, 391
anti-resonant 291, 387
cutoff 132
cyclical (Hertz)............................... 16
definition 13, 391

harmonic.......... 11, 19, 152, 312, 392
measurement................................. 16
resonant 113, 287, 395
self-resonant 280, 288, 291, 295

G

Gauss, Carl Friedrich............26, 329, 363
Generator 337, 391
Ground bounce 272
Guard band 233, 391

H

Harmonic 11, 19, 129, 152, 165, 284, 312, 325, 392
distortion.................................... 149
Henry 25, 54, 392
Henry, Joseph 26

I

i^2R 87, 91
Impedance..........122, 289, 300, 383, 392, *See also* Transmission line
calculation 203, 266
characteristic............................... 179
common mode 261
differential 261
parallel....................................... 134
series... 133
transmission line................188, 315
Inductive coupling 222, 257
Inductor, inductance 25, 312, 392
and rise time 150
calculating 55
ground bounce 270
ground plane............................... 170
hydraulic analog 50
ignition coil................................... 70
L/R time constant 75
lead 278, 373
methods for increasing 341
mutual inductance...................... 340
parallel..................................53, 103
phase shift....... *See* Phase angle, shift

reactance 99
RL circuit 58
self inductance 313, 340
series 52, 103
voltage and current through 66
voltage spike 70

J

j, imaginary. 123, 146, 290, 361, 392, 393

K

Kirchhoff 77, 169
    1st Law 78, 109
    2nd Law 40, 78, 112
Kirchhoff, Gustav 26

L

L/R time constant ....*See* Time constant
LC circuit 104, 107, 115
    resonance 280, 283
    transmission line 178, 311
Loop area 161, 300, 393
    differential trace 257
    differential traces 257
    EMI ....*See* EMI

M

Maxwell, James Clerk 1, 26, 330
Maxwell's equations...........................330
Microstrip 393
    configuration 29
    crosstalk.......................................226
    differential impedance 266
    impedance formula 203
    propagation speed 30
    tuning...33
Motor 333, 393

N

Norton equivalent circuit 89
Norton, E.L..26

O

Odd mode 249, 256, 260,
    See also Differential mode
Ohm 393
    definition 25
    meter.. 334
    zero-ohm resistor 303
Ohm, George Simon............................ 26
Ohm's Law 3, 57, 77, 86, 130, 136,
    160, 258, 298, 393
    definition 39
    for admittance............................. 142
    for reactance 101
Overlapping planes 304

P

Period 14, 393
Periodic Table....................................... 4
Phase angle, shift 9, 59, 99, 351, 394
    admittance 143
    and Fourier's Theorem................. 18
    definition 124
    differential traces........................ 252
    impedance................................... 123
    impedance curve......................... 280
    parallel bypass capacitors 290
    phase shift simulator................... 351
    reactance....................................... 97
    resonance 290
    RLC simulation 145
    Square Wave Simulator.............. 327
    variable network analyzer........... 378
Picket fence 174
Pole and zero 118, 291
Power 25, 86, 91
Power curve... 92
Propagation speed...................28, 30, 394
    and wavelength............................. 35
    electromagnetic field 155
    microstrip 30
Propagation time.........................28, 394
    and critical length 181
    capacitive loading................188, 206
    microstrip 31
Proton 3

Index

Q
Q 136

R
Radian 16, 394
RC circuit
    charging and discharging 57, 72
    equalization 319
    phase relationship 121
RC filter 131
RC time constant .... See Time constant
Reactance 283, 394
    at resonance 111
    capacitive 98
    definition 97
    inductive 98, 99
    Ohm's Law for 101
    parallel combination 102
    phase shift 98
    relationship to impedance 122
    series combination 102
Reference plane .. 230, 267, 300, 308, 396
    control common mode 172
    control of crosstalk 227
    control of EMI 161, 170
    control of impedance 197
    microstrip 393
    signal return 153
    signal transition 165
    stripline 395
    trace configuration 29
Reflect, reflection .... 150, 176, 379, *See also* Transmission line
    and differential traces 260
    backward crosstalk 224
    critical length 181
    crosstalk termination 229
    design issues 196
    echo illustration 369
    on transmission line 180
    simulation .... *See* Transmission line simulation
    simulation. *See* Crosstalk, simulation
    stubs 200
    TDR 376
    termination 180
    VNA 376
Reflection coefficient 183, 211
Relative dielectric coefficient 28, 190, 389, 394
Resistor, resistance .... 25, 77, 395, *See also* ESR
    hydraulic analog 38
    parallel 41, 81
    series 40, 79
    terminating ...... *See* Transmission line termination
    terminating .... *See* Termination
    voltage and current through 63
Resonance 61, 111, 290, 395
    LC circuit 113
    phase shift 115
    Q 134
    RLC circuit 140
Resonant, self 280
Reverse voltage diode 70
Right angle corner 383
Right hand rule 156, 330, 395
Rise time ...... 68, 147, 150, 183, 226, 228, 256, 395, *See also* di/dt
    and bandwidth 21
    and crosstalk 222
    and frequency 14, 21
    and ground bounce 271
    and lower voltages 298
    and signal coupling 160
    critical length 181
    definition 14
    fall time 16
    harmonics 129
    reflection 177, 369
    stub length 201
RL circuit 58, 74
RLC circuit 134, 137, 351
    resonance 140
    simulation 145
RMS (root mean square) 12, 395

S

Signal
    return 153, 157, 159, 298, 343
    return and plane boundary 305
    return and slots 232
    return on different layers 165
    return on power plane 164
    return path 163
    return plane discontinuity 167
    trace geometry 197
Signal coupling .... *See* Coupling
Signal timing, tuning 32, 33
Sine wave 8, 11
    and Fourier's Theorem 18
Single ended 249
Skin effect 312, 395
Slot 162, 199, 232
Spark plug 70
Square wave 14, 19, 252
Square wave simulation 325
Stackup 307
Stripline 242
Stub 168, 200, 395
Susceptance 119, 395
Symbols 24

T

TDR (time domain reflectometer) 375, 396
Termination 229, 242, 396
Thevenin 89, 192, 211
Time constant 57, 74, 396
    L/R 75
    RC 74, 132, 194
Trace configurations 29, 265
Trace layer 165, 196
Transformer 338
Transmission line 178, 396
    impedance 201
    lossy 311
    simulation 184, 367
    termination ...... *See* Transmission line termination
Transmission line simulation
    basic 207
    branch 219
    placement of termination 217
    series termination 215
Transmission line termination 179, 190
    AC 194
    differential traces 258, 260
    diode 195
    parallel 192
    series 194
    Thevenin 192
Tuned trace 396

U

Upward sloping curve 280

V

VNA (vector network analyzer) . 375, 396
Volt 25, 47, 54, 64
Volta, Allesandro 26
Voltage 6, 8, 37, 63, 396
    gradient 299
    ideal source 37, 67, 77, 129, 351, 389
    meter 13, 334
    noise 150, 202
    phase .... *See* Phase angle, shift
    reference 276
    reflection .... *See* Reflect, reflection
    reverse ...... *See* Reverse voltage diode
    source 89
    spike 70
Voltage divider 84, 131, 170, 211, 215, 238

W

ω (angular frequency) .... *See* Frequency, angular
Wavelength 35, 174, 396

ABOUT THE AUTHOR

Douglas Brooks began his career as an electronic design engineer in the aerospace industry and later as an applications engineer in the semiconductor industry. He then switched to business and advanced from VP of Marketing to General Manager and then to President of his own manufacturing company. Doug received his BS/EE and MS/EE from Stanford and a PhD from the University of Washington. He has spent four years of his career teaching upper division and graduate-level courses at the university level. Doug started his third company, UltraCAD Design, Inc., a printed circuit board design service bureau, in 1992, and has written many articles and given many seminars on high-speed design issues since joining the PCB design community.

Doug and his wife of 35 years, Alice, make their home in Bellevue, Washington, where they have raised three now-adult children.

informIT

www.informit.com

YOUR GUIDE TO IT REFERENCE

Articles

Keep your edge with thousands of free articles, in-depth features, interviews, and IT reference recommendations – all written by experts you know and trust.

Online Books

Answers in an instant from **InformIT Online Book's** 600+ fully searchable on line books. Sign up now and get your first 14 days **free**.

POWERED BY
Safari

Catalog

Review online sample chapters, author biographies and customer rankings and choose exactly the right book from a selection of over 5,000 titles.

Wouldn't it be great

if the world's leading technical publishers joined forces to deliver their best tech books in a common digital reference platform?

They have. Introducing
**InformIT Online Books
powered by Safari.**

- **Specific answers to specific questions.**
InformIT Online Books' powerful search engine gives you relevance-ranked results in a matter of seconds.

- **Immediate results.**
With InformIt Online Books, you can select the book you want and view the chapter or section you need immediately.

- **Cut, paste and annotate.**
Paste code to save time and eliminate typographical errors. Make notes on the material you find useful and choose whether or not to share them with your work group.

- **Customized for your enterprise.**
Customize a library for you, your department or your entire organization. You only pay for what you need.

Get your first 14 days **FREE!**
InformIT Online Books is offering its members a 10 book subscription risk-free for 14 days. Visit **http://www.informit.com/onlinebooks** for details.

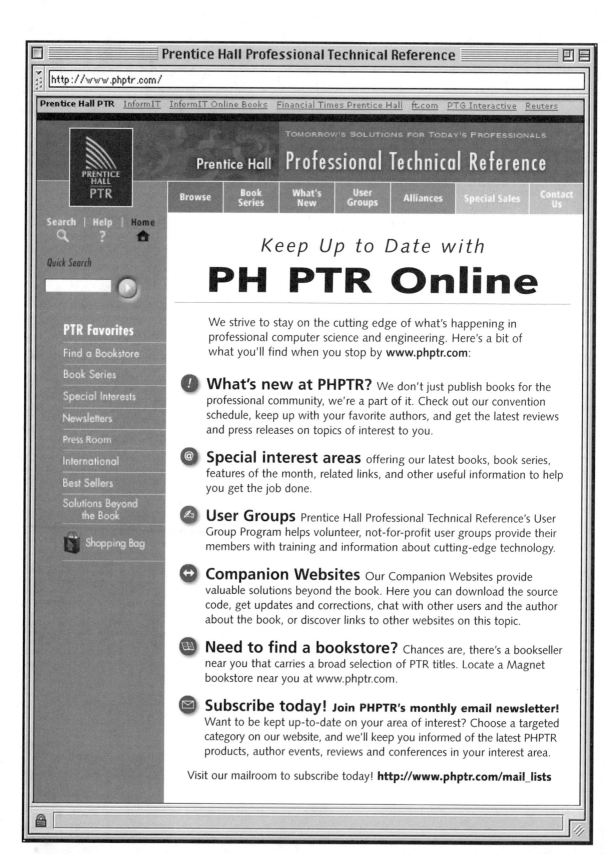